Spectral
and
Chemical Characterization
of
Organic Compounds

Spectral
and
Chemical Characterization
of
Organic Compounds

A Laboratory Handbook

THIRD EDITION WITHDRAWN

W. J. CRIDDLE and G. P. ELLIS

*School of Chemistry and Applied Chemistry,
University of Wales College of Cardiff,
Cardiff*

JOHN WILEY & SONS
Chichester · New York · Brisbane · Toronto · Singapore

Copyright © 1976, 1980, 1990 by John Wiley & Sons Ltd.

Reprinted September 1991
Reprinted May 1994

Library of Congress Cataloging-in-Publication Data:

Criddle, W. J.
  Spectral and chemical characterization of organic compounds : a
laboratory handbook / W. J. Criddle and G. P. Ellis.—3rd ed.
      p.   cm.
  ISBN 0 471 92715 5
  1. Chemistry, Analytic—Qualitative—Laboratory manuals.
2. Spectrum analysis—Laboratory manuals. 3. Chemistry, Organic–
–Laboratory manuals. I. Ellis, G. P. (Gwynn Pennant) II. Title.
QD271.4.C73   1990
547.3′4—dc20   89-21467
                                                           CIP

British Library Cataloguing in Publication Data:

Criddle, W. J. (William James), *1934–*
  Spectral and chemical characterization of organic
  compounds.—3rd ed.
  1. Organic compounds. Qualitative analysis & preparation
  of derivatives to confirm analysis
  I. Title   II. Ellis, G. P. (Gwynn Pennant), *1921–*
  547.34

ISBN 0 471 92715 5

Printed in Great Britain by Redwood Books, Trowbridge, Wiltshire

CONTENTS

Preface to first edition   vii

Preface to second edition   ix

Preface to third edition   xi

Safety in the laboratory   xiii

1. Preliminary tests   1

    1. Elemental analysis   1
    2. Ignition   3
    3. Colour and odour   3
    4. Determination of physical constants   3

2. Chemical and spectroscopic characterization of functional groups   7

3. Chromatographic methods   35

4. Chemical separation of organic mixtures   39

5. Preparation of derivatives   42

6. Tables of organic compounds and their derivatives   63

    Table 1.   Acetals   64
    Table 2.   Alcohols (C, H and O)   64
    Table 3.   Alcohols (C, H, O and halogen or N)   66
    Table 4.   Aldehydes (C, H and O)   67
    Table 5.   Aldehydes (C, H, O and halogen or N)   69
    Table 6.   Amides (primary), imides, ureas, thioureas and guanidines   69
    Table 7.   Amides, N-substituted   71
    Table 8.   Amines, primary aliphatic   71
    Table 9.   Amines, primary aromatic (C, H, (O) and N)   72
    Table 10. Amines, primary aromatic (C, H, (O), N and halogen or S)   75
    Table 11. Amines, secondary   76
    Table 12. Amines, tertiary   77
    Table 13. Amino-acids   78
    Table 14. Azo, azoxy, nitroso and hydrazine compounds   79
    Table 15. Carbohydrates   80
    Table 16. Carboxylic acids (C, H and O), their acyl chlorides, anhydrides and nitriles   81
    Table 17. Carboxylic acids (C, H, O and halogen, N or S)   86
    Table 18. Enols   88
    Table 19. Esters, carboxylic   88
    Table 20. Esters, phosphoric   90
    Table 21. Ethers   91
    Table 22. Halides, alkyl mono-   92
    Table 23. Halides, alkyl poly-   94
    Table 24. Halides, aryl   95
    Table 25. Hydrocarbons   96
    Table 26. Ketones (C, H and O)   98

Table 27. Ketones (C, H, O and halogen or N)   101
Table 28. Nitriles   102
Table 29. Nitro-, halogenonitro-compounds and nitro-ethers   103
Table 30. Phenols (C, H and O)   105
Table 31. Phenols (C, H, O and halogen or N)   108
Table 32. Quinones   110
Table 33. Sulphonic acids and their derivatives   110
Table 34. Thioethers (sulphides)   111
Table 35. Thiols and thiophenols   112

7. Pharmaceutical compounds   113

Introduction   113
Table P1. Compounds containing C, H, (and O)   113
Table P2. Compounds containing C, H, N, (and O)   114
Table P3. Compounds containing C, H, halogen (and O)   114
Table P4. Compounds containing C, H, N halogen (and O)   114
Table P5. Compounds containing C, H, N, S (and O)   115
Table P6. Compounds containing C, H, N, S, halogen (and O)   115

Index   117

PREFACE TO FIRST EDITION

Most courses in organic chemistry require that students be familiar with the spectral properties of organic compounds and also be able to recognize compounds by virtue of their characteristic chemical reactions. By combining these two approaches and correlating the results, it is usually possible to decide on the structure of the compound and, during the course of the work, the student learns a great deal of organic chemistry in a relatively short time. It also introduces to the student the methods which are adopted in research when a compound of unknown structure is found in nature or is synthesized in the laboratory.

Characterization of organic compounds in this way requires that the student have conveniently at hand the spectral and chemical data which have been accumulated over the years and which are spread through several theoretical books. The purpose of this handbook is to bring together the information that the student is likely to need in the laboratory and to present it in the most convenient way so that the minimum of time is used in searching for relevant information. There are several good theoretical books available on spectroscopy and therefore some knowledge of them is assumed. Similarly, we assume that the basis of the chemical tests in Chapter 2 will have been obtained from lectures and books on organic chemistry.

Systematic names of organic compounds have, as far as possible, been used throughout the book. At last, there are real signs that the old trivial names are being superseded in schools and colleges. We have included many of these as alternative names. The scope of the book is widened by the inclusion of a section on the characterization of pharmaceutical compounds. This section (Chapter 6) will be of particular interest to students of pharmacy.

We wish to thank Dr. D. J. Bailey (Welsh School of Pharmacy) for his valuable assistance with Chapter 6 and Professor W. H. Hunter (Chelsea College, University of London) for useful comments. We also thank the publishers for their patience and co-operation and Mrs. P. Bevan and Mrs. J. M. Watkins for secretarial assistance.

W. J. CRIDDLE
G. P. ELLIS

PREFACE TO SECOND EDITION

The warm reception given to our book has meant that improvements can be made in this edition. Some of these result from suggestions made by users while others result from changes in teaching methods and the availability of chemicals. For example, brief notes on the chemical basis of the functional group tests are now included in Chapter 2, the section on the interpretation of spectra in Chapter 3 has been extended and the selection of compounds in tables in Chapter 5 has been modified by the inclusion of low-priced compounds which have recently become available and the deletion of others which have disappeared from the chemical catalogues. The list of melting points of derivatives has also been amplified.

We are grateful to sharp-eyed students and others who have pointed out minor misprints in the previous edition and we welcome further suggestions for improving the usefulness of the book. We thank the staff of our publishers for their whole-hearted support and co-operation.

W. J. CRIDDLE
G. P. ELLIS

PREFACE TO THIRD EDITION

In recent years, there has been an increasing trend to rely on instrumental techniques for the identification of organic compounds. However, we believe that the practical experience of exploring and observing the chemical reactions of a variety of such compounds is the most efficient way (in time and effort) for students to learn organic chemistry.

Amongst the improvements which have been included in this edition is a section on the use of chromatographic methods which are widely used for separating, identifying and purifying organic compounds. In the NMR spectroscopy section, the table of chemical shifts has been completely revised, and readers should find it considerably more informative. The section on mass spectrometry has been supplemented by inclusion of the spectrum and associated fragmentation pattern of a simple ketone. One of the most useful reagents (xanthydrol) for preparing derivatives of carboxamides and sulphonamides has now become too expensive for class use. Diphenylmethanol (benzhydrol) is a good alternative, and its use is described. Finally, several minor improvements have been made elsewhere in the book.

We thank reviewers for their encouragement, and our publishers for their continued support.

W. J. CRIDDLE
G. P. ELLIS

SAFETY IN THE LABORATORY

It is essential that students be made aware of the statutory requirements on safety in the laboratory; in particular, they should be familiar with equipment provided for safety purposes.

All chemicals, especially unidentified compounds, should be treated with caution. Therefore, they should not be allowed to come into contact with the skin and we recommend wearing plastic gloves wherever any doubt exists. The eyes should always be protected by wearing safety spectacles.

Reactions involving hazardous reagents such as sodium, thionyl chloride, chlorosulphonic acid, phosphorus pentachloride, bromine, nitric acid and acyl chlorides should be carried out in an efficient fume cupboard. Flammable and malodorous compounds should be similarly treated.

PRELIMINARY TESTS

1. ELEMENTAL ANALYSIS

The identification of the elements contained in an organic compound is a first and most important step in organic analysis and is best effected by using Lassaigne's test. Other fusion tests have been described but none is as universally applicable as Lassaigne's. In this test the organic compound is decomposed by fusion with sodium. The presence of the elements nitrogen, halogens, sulphur or phosphorus in the original compound is then determined by various tests on the product. If nitrogen is present in the compound, fusion with sodium converts it into sodium cyanide which may be identified by the reaction of the cyanide ion in solution. Halide, sulphide or phosphate ions in the product similarly indicate the presence of halogen, sulphur or phosphorus in the compound being analysed.

Lassaigne's test does not provide information on the presence of carbon, hydrogen or oxygen. The vast majority of organic compounds contain carbon and hydrogen and it is usually possible to identify a compound without testing specifically for oxygen although the ferrox test (described below) gives a positive test for the majority of oxygen-containing compounds.

Lassaigne's test

CAUTION: Metallic sodium must be handled with great care as it reacts violently with water and many other compounds. It must never be allowed to come into contact with the skin and protective goggles should be worn.

For solids. Place a piece of metallic sodium (about 2 mm cube) in an ignition tube and heat gently until it is molten. Add the organic compound (about 200 mg) and continue heating gently until the contents of the tube are solid. Then heat more strongly and maintain at red heat for 2 min. While the tube is still red hot, plunge it into a 50 cm³ beaker containing distilled water (15 cm³). Boil for 3–4 min and filter the solution. Use the filtrate (A) for the tests given below.

For liquids. Heat a piece of metallic sodium (about 2 mm cube) in an ignition tube until the tube is one-third full of sodium vapour. Introduce the liquid (0.2 cm³) dropwise into the tube using a dropping-tube. When all the liquid has been added, heat strongly for 2 min and plunge the red-hot tube into a 50 cm³ beaker containing distilled water (15 cm³). Boil for 3–4 min and use the filtrate A for the following tests.

Nitrogen

To filtrate A (1 cm³) add iron (II) sulphate solution (10%, 1 cm³) and a little 2 M sodium hydroxide solution until a heavy precipitate of iron (II) hydroxide is obtained. Boil this for 2 min, cool and acidify (test with litmus paper) with 2 M sulphuric acid. A precipitate or coloration varying from deep blue to green indicates the presence of *Nitrogen* in the organic compound.

Sulphur

To filtrate A (1 cm^3) add freshly prepared sodium nitroprusside solution. A pink to purple coloration indicates the presence of *Sulphur*. This coloration is sometimes transient.

Halogens

To filtrate A (1 cm^3) add excess of 2 M nitric acid (test with litmus) and if nitrogen or sulphur has been detected by the above tests, boil the solution for 5 min in a beaker (fume cupboard) to remove hydrogen cyanide or hydrogen sulphide. Boiling is not necessary if nitrogen and sulphur are absent. To the cool solution add silver nitrate solution. A white or yellow precipitate indicates the presence of one or more of *Chlorine, Bromine* or *Iodine* in the organic compound. If this test is positive, the identity of the halogen will be revealed by the following tests, but before these are attempted it is advisable to carry out a blank test on the reagents alone.

If it is known that only one halogen is present it may be identified as follows: to filtrate A acidified with dilute sulphuric acid, add chloroform (1 cm^3) and chlorine water or 1% sodium hypochlorite solution (2 drops). Shake well and allow the chloroform layer to separate. A brown coloration in the chloroform layer indicates *Bromine*, a purple coloration *Iodine*, while no change in the colour of the chloroform indicates *Chlorine*.

If more than one halogen may be present, the following series of tests should be carried out:

(*a*) To the filtrate A (2 cm^3) add an excess of 2 M nitric acid followed by 5% mercury (II) chloride solution (1 cm^3) (POISON). A yellow precipitate which changes to orange or red on standing for a few minutes indicates the presence of *Iodine*. If a high concentration of iodide ions is present in the solution, the precipitate is orange or red immediately.

(*b*) To filtrate A (2 cm^3) add an equal volume of dichromate oxidizing mixture and boil gently for 2 min. Test the vapours produced with a filter paper dipped in freshly prepared Schiff's reagent. A purple coloration indicates the presence of *Bromine*.

(*c*) To filtrate A add an excess of 2 M nitric acid and then silver nitrate solution. Filter off the precipitate and treat it with an excess of a solution consisting of four volumes of saturated ammonium carbonate solution and one volume of ammonia solution (0.88). If a precipitate remains, filter this off and acidify the filtrate with dilute nitric acid. A white precipitate indicates the presence of *Chlorine*. It should be noted that silver bromide is slightly soluble in the solution used above. A faint precipitate obtained during this test should therefore be ignored.

(*d*) Acidify a further portion of filtrate A (2 cm^3) with ethanoic acid; bring the solution to the boil and cool. Add a drop of this solution to a piece of filter paper dipped in zirconium-alizarin solution (1% ethanolic alizarin and 0.4% aqueous zirconium nitrate) and allow the paper to dry. A red to yellow colour change indicates the presence of *Fluorine*.

Phosphorus

Treat a further portion of filtrate A (2 cm^3) with concentrated nitric acid (0.5 cm^3) followed by a 5% solution of ammonium molybdate. Heat on a boiling-water bath for 2 minutes. A yellow precipitate indicates the presence of *Phosphorus*.

Ferrox test for oxygen

Grind together equal weights of potassium thiocyanate and an iron (III) salt. Place the mixture (about 100 mg) in a test tube and add the organic compound directly if a liquid or as a saturated solution in benzene or chloroform if a solid. A purple coloration in the organic layer indicates the presence of oxygen in the compound. This test is specific for oxygen only if nitrogen and sulphur are absent.

2. IGNITION

Place some of the organic compound (0.1 g) on a spatula or in an ignition tube and heat until ignition occurs. Remove from the flame and observe the ignition characteristics. A clear flame indicates an aliphatic compound while a smoky flame is characteristic of aromatic and some unsaturated compounds. Continue the ignition until no further change occurs; the presence of a residue shows that the original compound contains a metal atom and the residue should be examined by the standard inorganic procedures for the identification of metals. In the majority of cases, a flame test carried out on the residue, which should be acidified by addition of concentrated hydrochloric acid (1 drop), will be sufficient to identify the metal atom present. It is sometimes possible to recognize the odour of the vapours released during ignition; some compounds (for example, carbohydrates, aliphatic hydroxy acids and their salts) char readily while others (for example, benzoic acid) sublime.

3. COLOUR AND ODOUR

The majority of organic compounds are colourless when pure but some compounds become discoloured on standing due to the formation of small amounts of coloured impurities. If a pure compound is coloured, it will contain one or more chromophoric groups, for example, nitro, nitroso or azo group, or be a quinone or have an extended conjugated system of four or more double bonds. The nitro group on its own confers very little if any colour on a compound but if an auxochromic substituent such as a hydroxyl or amino group is also present, the very pale yellow colour is intensified. An indication of the colour of many compounds is given in the tables of melting points at the end of this book.

Some organic compounds have characteristic odours which can be used tentatively to guide us in organic analysis but since smell cannot often be described in words, the student's best approach should be to try to 'memorize' the smell of some common compounds.

4. DETERMINATION OF PHYSICAL CONSTANTS

Melting point

Before a melting point is determined, the sample must be pure and free of solvent. Seal off a standard melting-point tube at one end in a Bunsen flame. Introduce the sample to a depth of about 2 mm at the sealed end of the tube (rubbing the tube with a nail file often facilitates this). Place the tube in an electrically heated melting-point apparatus. Adjust the rate of heating so that the temperature rises about $3-4°C$ min^{-1}. The temperature at which a meniscus forms from the molten sample is the required melting point.

Notes: (*a*) It is often better to determine the approximate melting point of the sample first and then redetermine the melting point more accurately on a second sample by slowing down the heating rate to $2°$ min^{-1} as the approximate melting point is approached.

(*b*) For samples which are thermally unstable it is preferable to determine the approximate melting point in the ordinary way and then to determine the accurate value by heating the oil-bath to within $10°$ of the rough value before inserting the sample; the temperature is then raised at $2° \text{ min}^{-1}$ until the compound melts. This reduces the period during which the sample is being heated and thus minimizes thermal decomposition.

Mixed melting point

When two different compounds are mixed together and the melting point of the mixture is determined, it is found that melting begins at a temperature several degrees below that of the lower-melting pure compound. This technique of *mixed melting points* may therefore be used to determine whether or not the two samples are identical. A depression in the melting point of a sample when mixed with another indicates that the compounds are different. The correct procedure for such a determination is as follows: grind together equal weights of the known and unknown materials and introduce the mixture into a melting-point tube. Place a little of each of the two pure compounds in two other tubes and determine the melting points of all three simultaneously. If the mixture melts more than $5°$ below the melting point of either of the pure samples, the latter are different. If they are identical, the three samples will melt at the same temperature.

Boiling point

Siwoloboff's method

Take a glass tube about 5 cm long and 0.5 cm internal diameter and a standard melting-point tube and seal each at one end only. Introduce the liquid under examination (0.5 cm^3) into the larger tube and place in it the melting-point tube, *open end* in the liquid. Attach the tubes to a short-immersion $360°$ thermometer with the liquid at the same level as the thermometer bulb. Immerse the thermometer in a liquid paraffin bath to a depth of 3 cm. Heat the bath at a constant rate with continuous stirring until a rapid stream of bubbles emerges from the lower end of the smaller tube. Stop heating at this point and note the temperature at which the liquid rises rapidly into the smaller tube. This is the boiling point of the liquid. If the sample is impure (for example, contains a small amount of water) Siwoloboff's method will give misleading values. It is then better to remove the impurity by fractional distillation or by thorough drying with a desiccant.

Solubility in various solvents

The solubility of an organic compound in water, ether, 2 M hydrochloric acid and 2 M sodium hydroxide can often furnish useful information about the nature of the compound. However, the presence of more than one functional group may have such a profound effect on the solubility that it is often impossible to make deductions about the functional groups present from solubility data. For example, 1,3-dihydroxybenzene is extremely soluble in water but the introduction of a butyl group into the 4-position gives a compound which is only slightly soluble. Even positional isomers sometimes differ greatly in their solubility, for example, the solubilities of 1,2-, 1,3- and 1,4-dihydroxybenzene in water at $20°$ are 45, 210 and 7% respectively. The following table of solubilities should therefore be used with caution, and it is

4

likely to be most accurate for monofunctional compounds. It gives an indication of the solubility of various types of organic compounds in ether, water, 2 M hydrochloric acid and 2 M sodium hydroxide. The compounds are grouped together according to the elements which are identified in the Lassaigne test. A + sign means that the compounds of that class have a solubility in the particular solvent in excess of 5%. A lower solubility is indicated by a − sign. A class of compounds whose members vary greatly in solubility in a particular solvent is shown as ±.

Table of solubilities

	Ether	Water	2 M HCl	2 M NaOH	Comments
Compounds containing C, H, O, Metal					
Carboxylic acids					
aliphatic	+	+[a]	+[a]	+	[a]Insol. if >4C atoms.
aromatic	+	−	−	+	
metal salts	−	+	±	+	
Phenols					
monohydric	+	−[b]	−	+	[b]Simple phenols are sol.
di- and tri-hydric	+	+[c]	±	+	[c]1,3,5-trihydroxybenzene is insol.
phenoxides	−	+	±	+	
Aldehydes and Ketones					
aliphatic	+	−[d]	−[d]	−[d]	[d]Sol. if <4C atoms.
aromatic	+	−	−	−	
Acetals	+	±	−[e]	±	[e]Hydrolysis occurs
Alcohols	+	+[f]	+[f]	+[f]	[f]Sol. if <4C atoms.
Carbohydrates	−	+	+	+	
Polyols	−	+	+	+	
Esters	+	−[g]	−[h]	−[h]	[g]Sol. if <4C atoms. [h]May hydrolyse to sol. products
Anhydrides	+	±[i]	−	+	[i]Aliphatic, +; aromatic, −.
Lactones	+	−[j]	−	+	[j]γ-Butyrolactone is sol.
Quinones	+	−	−	+	
Ethers	+	−	−	−	
Hydrocarbons	+	−	−	−	
Compounds containing C, H, N(O)					
Amines					
pri. aliphatic	+	+	+	−	
s-aliphatic	+	−[k]	+	−[k]	[k]Sol. if <4C atoms.
t-aliphatic	+	±[l]	+	−	[l]Variable.
pri. aromatic	+	−	+	−	
s-aromatic	+	−	+	−	
t-aromatic	+	−	+	−	
Amides	−	−[m]	−	−	[m]Sol. if <6C atoms.
N-Substituted amides	+	−	−	−	
Imides	−	−	−	+	
Ammonium salts	−	+	+[n]	+	[n]Depends on solubility of free acid.
Nitrocompounds	+	−	−	−	
Amino-acids	−	±[o]	±[o]	±[o]	[o]Variable.
Arylhydrazines	+	−	+	−	

5

Table of solubilities (*cont.*)

	Ether	Water	2 M HCl	2 M NaOH	Comments
Compounds containing C, H, S(O)					
Sulphonic acids	−	+	+	+	
Thiols and thiophenols	+	−	−	+	
Compounds containing C, H, P(O)					
Phosphate esters	+	−	−	−	
Compounds containing C, H, Halogen(O)					
Alkyl and aryl halides	+	−	−	−	
Acyl halides	+	+q	+q	+q	qDecomposed, alkyl compounds rapidly.
Compounds containing C, H, Halogen, N(O)					
Quaternary ammonium salts	−	+	+	+	
Hydrohalides of organic bases	−	+	+	+	
Compounds containing C, H, Halogen, S(O)					
Sulphonyl halides	+	−	−	−	
Compounds containing C, H, N, S(O)					
Thioamides	−r	−r	−	+	rSol. if <3C atoms.
Sulphates of organic bases	−	+	+	−	
Sulphonamides	+	−	−	+	

CHEMICAL AND SPECTROSCOPIC CHARACTERIZATION OF FUNCTIONAL GROUPS

When the elements which are present in an organic compound have been determined, it is then necessary to ascertain how these are arranged in the molecule, that is, what functional group(s) it contains. For this purpose, use is made of the chemical reactions and spectral data which are characteristic of each function.

CHEMICAL TESTS

The tests described below are grouped according to the element(s) detected in the Lassaigne test. Compounds containing only C, H, O should be examined by the tests given in Table I. If one of the elements halogen, N, P or S is present, the tests in Tables II–V should take precedence. Where two or more of these elements are present, the initial tests should be for the composite group containing the relevant elements (Tables VI–IX); for example, if N and S have been detected, the presence of a sulphonamide group ($-SO_2 \cdot NH_2$) among others should be investigated as described in Table VIII. If these tests are negative, a search should be made for the presence of groups containing the individual elements, that is, sulphur-containing functions (Table IV) and nitrogen-containing functions (Table II).

When a functional group has been identified, it should be possible to correlate this with the results of the preliminary tests so that examination of the appropriate melting point table (p. 59) will narrow the choice to one or more possible compounds. In such cases, consideration of the structure of the likely compounds may suggest further functional group tests to differentiate between them and thus enable the compound to be identified. For example, an organic compound, m.p. 144°, containing C, H, N, O was shown to be a carboxylic acid. Reference to Table 17 shows that it may be one of two compounds having a melting point of 144°, namely, 2-hydroxy-3-nitrobenzoic acid or 2-aminobenzoic acid. These may be distinguished by the appropriate tests for the other functional group, that is, phenolic, nitro and amino group respectively. The identification should then be confirmed by the preparation of one or more of the derivatives given in Table 17.

A series of tests contained within heavy horizontal lines is for related functions. In many cases, if the first test in a series is negative, the subsequent tests in that particular series may be omitted.

INFRARED SPECTRAL DATA

Preparation of samples

(a) Solids. Solid organic compounds are normally treated in one of three ways. Firstly, by making a solution (~10% m/v) using an anhydrous solvent which interferes as little as possible with the spectrum of the compound. Solvents commonly used are carbon tetrachloride, trichloromethane (chloroform), carbon disulphide and hexane. Note that it is important that the solvent does not associate in any way with the substrate. Spectra may then be obtained using standard path length cells (0.025–1.0 mm). Dual-cell operation, i.e. with solvent alone in the reference beam, is sometimes useful in minimizing the effects of solvent absorption, but it should be noted that this technique will not allow absorption of substrate to be measured where strong solvent absorption occurs. Secondly, the solid ($\sim$10 mg) may be mixed with anhydrous potassium bromide (100–200 mg, preferably fused immediately prior to use) and the mixture ground to a fine powder in either an agate micromill or using an agate pestle and mortar. The powder obtained is transfered to a special evacuable die and press, and subjected to a pressure of up to 20 tons per sq. in. when the mixture is converted to a transparent disc* suitable for the specially designed holders. This method is particularly recommended as the potassium bromide is transparent over a wide spectral range and dual-beam operations are not normally necessary.

Finally, the solid may be converted to a mull, i.e. ground to a paste with a non-volatile oil, e.g. Nujol (a high-boiling paraffin hydrocarbon) or Flurolube (a mixture of high-boiling fluorinated hydrocarbons). Use of these two oils allows most of the usual infrared region to be scanned. The mull may be conveniently prepared using a standard B14 cone and socket joint by taking the cone and adding Nujol (1 drop) and the sample (2–5 mg) to the ground area. The cone is inserted into the socket and the two parts twisted. Mull formation is rapid and the mull can be removed from the joints using a small spatula. The mull is placed onto a rock-salt plate,* covered with a second plate and pressed into a thin film.

(b) Liquids. Organic liquids may be examined either as solutions (as for solids) or as films. For volatile liquids (b.p. < 120°) a closed cell should be used. Less volatile liquids may be examined as described for mulls, one drop of liquid usually being sufficient.

Interpretation of spectra

(a) Most functional groups have their characteristic absorptions (stretching frequencies) in the region 4000–1400 cm^{-1}. This region should be examined first when a knowledge of the elemental constituents of the unknown compound has been established.

(b) When classifying the possible functional groups according to group frequencies, it is important to keep in mind the physical state of the compound when the spectrum was obtained. Spectra of solids and liquids usually show a lowering in the group frequencies of polar groups due to hydrogen bonding. The lowering of group frequencies as a result of molecular association (intermolecular bonding) is often accompanied by a broadening of the peak. Spectra obtained in dilute solution or in the vapour phase do not show these effects unless they result from intramolecular bonding.

(c) Absorptions found in the region 1400–900 cm^{-1} (fingerprint region) are normally more difficult to interpret due to the complexity of group vibrations in this region. Many vibrational (bending and stretching) modes are possible in this region and assignments should be made with caution.

(d) The region below 900 cm^{-1} is mainly useful for providing information regarding aromatic substitution patterns (Table XI) and for some stretching frequencies of carbon—halogen groups.

---

*Note that when handling rock salt or potassium bromide plates, discs or cells, protective (rubber or PVC) gloves should be worn to prevent fogging.

(e) Abbreviations used for infrared intensities are: s, strong; m, medium; w, weak; v, variable; ν, stretching frequency; γ, out-of-plane bending frequency; δ, in-plane-bending frequency.

The information given in Tables I–IX includes infrared spectral data which will be adequate for confirming the presence of most of the functional groups normally encountered by undergraduates. In addition, further information of carbon–hydrogen vibrations is included in Tables X and XI.

It should be noted that the type of spectrometer used may have a marked effect on the infrared spectrum. Instruments employing rock-salt prisms do not have the resolving power of those using diffraction gratings and consequently the infrared spectra obtained in such instruments often have fewer peaks than those obtained in the latter type. To quote a simple example, a compound of the type $CH_3(CH_2)_nCH_3$ will show only two CH stretching bands at 2944 and 2865 cm^{-1} when examined in an instrument with rock-salt optics. However, a grating instrument will show four bands at 2962, 2926, 2872 and 2853 cm^{-1}, i.e. the symmetric and asymmetric methyl and methylene stretching modes.

ULTRAVIOLET SPECTRAL DATA

Preparation of samples

Both solid and liquid compounds are usually examined as solutions in a solvent which has little or no absorption in the region 220–800 nm. The concentration of the solution will depend on the absorbance value(s) at whatever absorption maxima are observable. Solvents should be of spectroscopic quality and among those commonly used are water, methanol, ethanol, chloroform, carbon tetrachloride and hexane. Cells used are either glass or silica (for absorptions below 340 nm) and usually have a 1 cm path length. Dual-cell operation as described for infrared spectroscopy is usual.

Interpretation of spectra

Many groups which absorb in the ultraviolet show maxima over a wide wavelength range and it is therefore difficult to be specific about any particular formation. However, high absorbance values are an indication that a conjugated system is present and may be confirmed by other methods. Ultraviolet data are usually presented by quoting the molar extinction coefficient (ε_{max}) at the appropriate wavelength (λ_{max}). Sometimes, particularly in pharmaceutical literature, the specific extinction coefficient ($E_{1\,cm}^{1\%}$) is quoted and is sometimes useful for characterizing a compound in much the same way as a melting-point.

Useful information regarding the structure of an unknown organic compound containing conjugated polyene and $\alpha\beta$-unsaturated carbonyl systems may be obtained from Woodward's rules which allow reasonably accurate values of λ_{max} to be calculated for a given structure. These values often allow possible structures to be differentiated although it should be noted that application of these values alone will not provide an identification of an unknown compound.

NUCLEAR MAGNETIC RESONANCE SPECTRAL DATA

Preparation of samples

The compound should be dissolved in a solvent which itself does not absorb. Carbon tetrachloride, carbon disulphide and deuterotrichloromethane (deuterochloroform, $CDCl_3$) are commonly used. A concentration of about 10% w/v is desirable but it is often possible to work with slightly lower concentrations by adjusting the spectrometer. It is important to use a thoroughly dry solvent and to exclude moisture and water vapour until the spectrum has been obtained. For some compounds, water, deuterated dimethyl sulphoxide, deuterium oxide or dilute hydrochloric acid may be used as solvent but a part of the spectrum will be obscured by the strong solvent absorption (except for deuterated solvents).

A small amount of tetramethylsilane should be added to the solution to serve as an internal standard but since this compound is insoluble in aqueous solutions, it should be replaced by sodium 3-(trimethylsilyl)propanesulphonate.

Interpretation of spectra

The following characteristics should be considered when interpreting the spectrum of a hydrocarbon or any organic compound containing C–H bonds:

(a) The number of protons which give rise to each signal may be deduced by measuring the fall in the integrator trace for each signal. If the molecular formula is not known, it may only be possible to determine the ratio of the number of protons in the various signals.

(b) The chemical shift of the signal should be noted and compared with that of hydrogens in known environments, for example, as listed in Tables XII and XIII.

(c) The multiplicity of the signal provides information about the number of hydrogen atoms which are attached to adjacent atoms. This valuable feature applies mostly to aliphatic groups where first-order spin–spin coupling occurs (i.e. the $(n + 1)$ rule is valid) but may also provide information on the orientation of hydrogens on an aromatic ring.

(d) Replacement of a hydrogen by deuterium on treating the compound with D_2O is accompanied by the disappearance of a signal and indicates the presence in the compound of a hydrogen attached to oxygen, nitrogen or sulphur.

(e) Most hydrogens appear as sharp signals but a hydrogen attached to oxygen sometimes gives a more diffuse peak, especially if hydrogen exchange is catalysed by a trace of acidic impurity. The hydrogen atoms of a primary amide ($CONH_2$) appear as an ill-defined hump but the hydrogen of secondary amides (CONHR) usually gives a sharper signal.

(f) The type of absorption signal(s) exhibited by hydrogens attached to benzene or other aromatic rings depends on the electronic character of substituent(s). A methyl group has a very small effect on the magnetic environment of the five ring-hydrogens in toluene which appear as a singlet at $\delta\,7.20$ (compared with $\delta\,7.37$ for benzene hydrogens). Substituents (such as methoxy) which are electron-releasing have two effects on the signals of the aromatic hydrogens: (a) they all suffer a diamagnetic (upfield) shift, and (b) this shift is usually more pronounced on the *ortho* than on the *meta* and *para* hydrogens. Consequently, in contrast to the simple signal for those of toluene, the signal of the aromatic hydrogens in methoxybenzene is a complex multiplet at $\delta\,7.4$–6.7. An electron-withdrawing substituent exerts a general paramagnetic (downfield) effect which is more strongly felt by the *ortho*-placed hydrogens; for example, the five hydrogen atoms of nitrobenzene absorb at $\delta\,8.3$–7.1.

(g) Aromatic hydrogens which are in different magnetic environments undergo spin–spin coupling but the first-order splitting rule does not apply because the signals are too close together. Provided that not more than two hydrogens are *ortho* to one another, it is often possible to interpret the spectrum and assign the signals to individual atoms. A tetrasubstituted

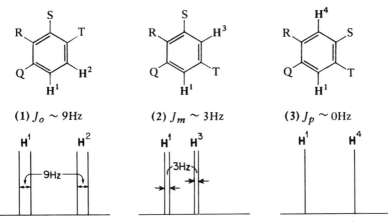

Figure 1. *Ortho, meta* and *para* hydrogens and their typical signals

benzene usually gives a simple signal pattern in the δ 6.5–9.5 region, provided all the substituents are not magnetically identical. The relative positions of the remaining two hydrogens in such compounds may be determined by observing the extent of spin–spin coupling as measured by the coupling constant, J, typical values of which are shown in formulae (1)–(3). Characteristic signals for these compounds are shown in Figure 1. Other coupling constants are given in Table XIV.

The difference in chemical shift between H^1 and H^2, H^1 and H^3, and H^1 and H^4 depends on the magnetic environment created by the substituents Q, R, S and T; the smaller this difference, the nearer together the signals will be.

(h) The spectrum of trisubstituted benzenes can sometimes be analysed, especially if the signals are well separated. For example, the three aromatic hydrogens of 4-methyl-2-nitrophenol (4) are easily identified in Figure 2. The signal of H^5 shows spin–spin coupling with H^3 superimposed on that with H^6.

(i) Of the disubstituted benzenes, the 1,4-isomers give the most easily recognized signal. When both substituents are identical, all the ring hydrogens are in similar positions, they have the same chemical shift and so appear as a singlet. Examples are 1,4-dimethyl- and 1,4-dibromo-benzenes. Compounds containing *different* substituents at C-1 and C-4 also have a degree of symmetry which is reflected in the spectrum. For example, 4-methoxybenzoic acid (5) has two pairs of hydrogens, each pair having a similar chemical shift. Although each doublet consists mainly of a tall and a short peak, there are other minor peaks close to them. The tall peaks are usually flanked by the shorter ones and the distance between the two doublets is a function of the different magnetic character of the substituents. A discussion of the carboxyl hydrogen signal in Figure 3 is given in the next paragraph.

(j) Hydrogen bonding results in a paramagnetic shift of the hydrogen signal. This is sometimes sufficient to move the signal beyond δ 10 and in such cases that part of the spectrum (which is referred to as an offset) is shown on a higher baseline than the main part. Most carboxylic acids exist in solution as dimers held together by intermolecular hydrogen bonding (6). Intramolecular hydrogen bonding is common in compounds such as 2-hydroxyacetophenone, 2-nitroaniline, and 1,3-diketones (7) which tautomerize to enols (8) stabilized by hydrogen bonding. From the spectrum of the equilibrium mixture of (7) and (8), it is possible to estimate the percentage of each tautomer present.

11

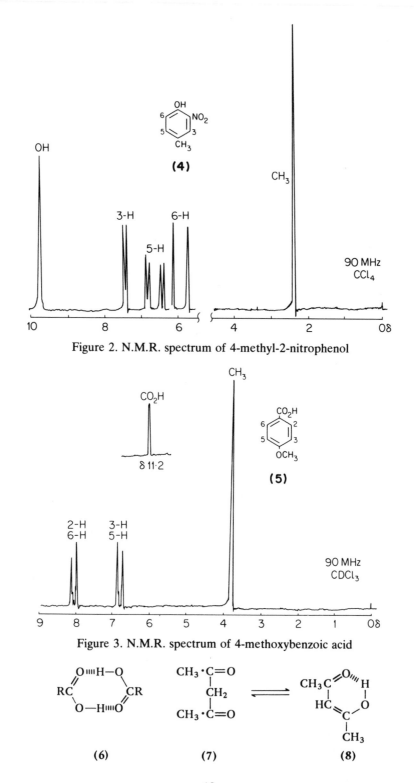

Figure 2. N.M.R. spectrum of 4-methyl-2-nitrophenol

Figure 3. N.M.R. spectrum of 4-methoxybenzoic acid

(6)   (7)   (8)

12

MASS SPECTROMETRIC DATA

When a molecule is bombarded with a beam of electrons, it loses an electron and its positive ion (molecular ion) is formed. If the voltage is sufficiently high, the molecular ion breaks down into fragments some of which are positively charged ions while others are neutral species (molecules or radicals). The positive ions are separated in the spectrometer by magnetic and electric fields and the mass to charge ratio (m/z) of each ion is recorded by the instrument. The charge on the majority of ions is unity and the spectrum thus records the mass of each ion. Modern mass spectrometers measure this value very accurately and this enables the identity and the number of atoms in each ion to be determined. Since the molecular ion has effectively the same mass as the parent molecule, the accurate molecular weight of the compound may be obtained in this way, e.g. both ethanal, CH_3CHO and propane, $CH_3CH_2CH_3$, have a molecular weight of 44 but an accurate determination distinguishes ethanal (44.0261) from propane (44.1248). The identity of many of the fragmentary ions may be similarly deduced and this knowledge contributes to the determination of the structure of the compound. The most abundant fragment ions from a large number of compounds are tabulated in *Compilation of Mass Spectral Data* by A. Cornu and R. Massot (Heyden and Son, London), and *Eight Peak Index of Mass Spectra*, 3rd edition (Royal Society of Chemistry, Cambridge).

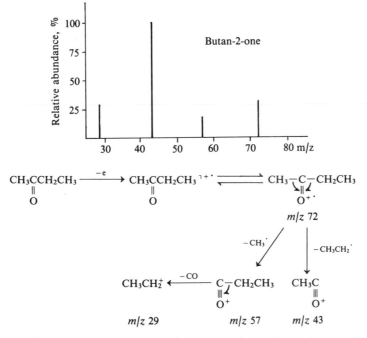

Figure 4. Mass spectrum and fragmentation of butan-2-one

Figure 4 shows how a simple ketone (butan-2-one) fragments in the mass spectrometer (only the main fragments are shown). The peak at m/z 72 is the 'molecular ion' which represents the molecular mass of the compound. The high intensity of the m/z 43 ion shows that it is relatively more stable than the other ions. A symbol, ⌢↘, is used in the fragmentation scheme to indicate the transfer of *one* electron from a two-electron bond.

Table I. Functional Groups Containing C, H, O

Chemical test	Observation	Functional group	Spectral data
1. Dissolve the organic compound (0.2 g) in water or aqueous ethanol (5 cm³) and add M sodium hydroxide solution (1 drop). Test the solution with BDH '1014' indicator (2 or 3 drops).	Green coloration	**Carboxyl (RCOOH)** **Phenol (ArOH)** **Enol (RC=CHR′)** $\overset{\mid}{O}$H	**Carboxyl.** I.R. 3000–2500 cm⁻¹, R = Ar and Alk, ν O—H (broad and complex); 1720–1710 cm⁻¹, R = Alk, ν C=O(m) (lowered by $\alpha\beta$-unsaturation); 1700–1680 cm⁻¹, R = Ar. ν C=O(m); 900–860 cm⁻¹, (m, broad), γ O—H. N.M.R. δ 10.0–13.0, unaffected by dilution
a. Prepare a saturated solution of the compound in 50% aqueous ethanol and add an equal volume of 5% sodium bicarbonate solution	Carbon dioxide evolved	Carboxyl	**Phenol.** I.R. 3620–3610 cm⁻¹, O—H (w, very sharp—no hydrogen bonding); 3150–3050 cm⁻¹, ν O—H (s, broad—normal absorption for hydrogen-bonded phenols); 1410–1310 cm⁻¹, δ O—H (m, broad); 1230 cm⁻¹, ν C—O (s, broad); ~650 cm⁻¹, γ O—H (variable). N.M.R. δ 4.5–6.8, varies with concentration. If hydrogen bonded, δ 8.0–13.0
b. (i) Treat with freshly prepared aqueous neutral iron(III) chloride (1 or 2 drops) (ii) Treat with 5% methanolic anhydrous iron(III) chloride	Wide range of colorations (See Tables)	**Phenol or enol**	**Enol.** I.R. Similar to phenols, but with differences due to presence of alkenic C=C and (usually) absence of characteristic aromatic absorption (see hydrocarbon absorption below). N.M.R. δ 14.0–17.0 for O—H stabilized by hydrogen bonding N.B. The O—H → O—D exchange is facile in D₂O for all O—H groups
c. If Test 1b is positive, prepare a cold solution of mercury(I) nitrate in 2 M nitric acid and add the organic compound	Immediate grey precipitate of mercury	Enol	

Notes

1. Below pH 10, BDH 1014 indicator is green. Addition of a small amount of dilute alkali will be insufficient to neutralize the given amount of an organic acid and the coloration with the indicator will therefore be green. If the organic compound is not acidic, or perhaps contains only a small amount of an acidic impurity, addition of the alkali will be sufficient to cause an increase in pH, giving a pink coloration.

a. $RCO_2H + NaHCO_3 \rightarrow RCO_2Na + CO_2 + H_2O$
   Phenols and enols are not normally sufficiently acidic to give the above reaction. Exceptions occur when phenols are extensively substituted with strong electron attracting groups, e.g., 2,4,6-trinitrophenol.

b. Phenols and enols complex with iron(III) to give a wide range of colorations. Note, however, that this is not always so and, in some instances, the coloration is extremely transient.

c. Enols reduce mercury(I) compounds to metallic mercury which appears as a grey precipitate.

	Aldehyde (RCHO) Ketone (RCOR')	

2. Treat the organic compound with 2,4-dinitrophenylhydrazine in either 5 M hydrochloric acid (for water-soluble compounds) or ethanolic phosphoric acid. N.B. Acetals are hydrolysed in acid solutions to aldehydes and will give a positive test with 2,4-dinitrophenyl-hydrazine. Silver mirror tests will be negative unless the acetal is first hydrolysed.

Yellow to red precipitate

Aldehyde. I.R. 2720 cm^{-1}, aldehyde ν C—H(w); 1725–1715 cm^{-1}, R = Alk, ν C=O (s, lowered by $\alpha\beta$-unsaturation to ~1685 cm^{-1}); 1700 cm^{-1}, R = Ar, ν C=O (s)
N.M.R. δ 9.0–10.0

Ketones. I.R. 1720–1710 cm^{-1}, R and R' = Alk, ν C=O (s, affected by $\alpha\beta$-unsaturation as for aldehydes); 1690 cm^{-1}, R = Alk, R' = Ar ν C=O (s); 1665 cm^{-1}, R and R' = Ar, ν C=O (s)
N.M.R. See Table XII for effect of keto group on chemical shift of nearby protons

N.B. Values for ν C=O in the I.R. spectra of aldehydes and ketones will be reduced if the group is concerned in hydrogen bonding, particularly with alcohol and phenol groups

a. Add 2 M ammonium hydroxide dropwise to 5% silver nitrate solution until the precipitate just dissolves. Add the organic compound and warm on a water bath. A little ethanol may be added if the compound is not water-soluble.

Aliphatic **aldehydes**

Silver mirror formed

b. If Test 2a is negative, treat 5% silver nitrate with 2 M sodium hydroxide (2 drops). Dissolve the precipitate obtained in the minimum of 2 M ammonium hydroxide. Add the organic compound and proceed as in Test 2a

Aromatic **aldehyde**

Silver mirror formed

c. Tests 2a and 2b

No silver mirror

Ketone

Notes

RCHO + O$_2$N ⟶ (2,4-dinitrophenylhydrazone)

yellow–red

R'R''CO + O$_2$N ⟶ (2,4-dinitrophenylhydrazone)

yellow–red

a. Ammoniacal silver nitrate is a sufficiently strong oxidizing agent to oxidize aliphatic aldehydes only.
b. The silver solution prepared as in 2b above is a moderately strong oxidizing agent of both aliphatic and aromatic aldehydes.
c. Ketones do not normally give silver mirrors but there are a few exceptions, e.g., acetophenone and cyclohexanone.

Table I. Functional Groups Containing C, H, O (*cont.*)

Chemical test	Observation	Functional group	Spectral data
3. (i) Add 5% chromium trioxide in 2 M sulphuric acid (3 drops). Warm at 40–50° for 1 min	Red colour changes to green	**Alcohol (AlkOH)**	I.R. ~3300 cm^{-1}, v O—H (s, broad for most alcohols in solid or liquid state). If in dilute non-polar solution, v O—H can be up to 3620 cm^{-1} (no hydrogen bonding)
(ii) Dissolve the compound in water, dioxan or propanone (acetone) and add 40% cerium(IV) nitrate in 2 M nitric acid (3 or 4 drops)	Red coloration		N.M.R. δ 1.0–5.0 depending on solvent and concentration

Note

3. (i) and (ii) Redox reactions involving valency changes of metal ions result in colour changes of the reactant:

Cr^{3+} (orange-red) $\rightarrow$ Cr^{2+} (green)

Ce^{4+} (yellow-orange) $\rightarrow$ Ce^{3+} (red)

Chemical test	Observation	Functional group	Spectral data
4. If Test 3 is positive, and the unknown compound is a solid, dissolve in water and add 10% ethanolic 1-naphthol followed by slow addition (down side of test tube) of concentrated sulphuric acid	Violet coloration at liquid interface	**Carbohydrate**	I.R. Similar in many respects to the spectra of alcohols. Also 930–804 cm^{-1}, asym. ring vibration; 898–884 cm^{-1}, anomeric equatorial C—H deformation (β-sugars); 888–872 cm^{-1}, equatorial C—H deformation; 852–836 cm^{-1}, anomeric axial C—H deformation (α-sugars)
			N.M.R. Complex signal δ 3.0–5.0 in D$_2$O
a. If Test 4 is positive, prepare a solution containing equal volumes of 0.1 M copper(II) sulphate and 0.1 M sodium potassium tartrate. Add 0.1 M sodium hydroxide until the precipitate just dissolves and then add the organic compound. Warm on a boiling-water bath for up to 5 min (Fehling's test).	Brick-red precipitate	**Reducing carbohydrate**	
b. Dissolve the compound in water and add a solution of 5% copper(II) ethanoate (acetate) in 1% aqueous ethanoic (acetic) acid. Boil gently for up to 2 min	Brick-red precipitate	**Monosaccharide**	

c. If Test 4b is positive, take the compound (0.1 g) and 1,3,5-trihydroxybenzene (phloroglucinol, 0.1 g) and dissolve in 2 M hydrochloric acid (2 cm³). Heat to boiling for up to 2 min

Hexose — Pale yellow
Pentose — Intense red-brown coloration

Note

4. The actual reaction pathway is uncertain, but the most likely explanation involves the formation of a naphthoquinone of the general type

a. Reducing sugars will reduce Cu(II) tartrate solutions to give copper(I) oxide (brick-red) as the product.
b. As a. above.
c. As 4. above.

5. (i) Mix equal volumes of saturated methanolic hydroxylamine hydrochloride and saturated methanolic potassium hydroxide. Add the organic compound and heat on a boiling-water bath. Cool, acidify with 0.2 M hydrochloric acid and add 5% aqueous iron(III) chloride (3 drops). N.B. This test should not be applied if Tests 1b and 1c are positive

(ii) Dissolve the compound in ethanol; add methanolic potassium hydroxide (1 drop) and phenolphthalein (1 drop). Prepare a similar mixture but omit the compound under examination. Heat both samples in a boiling-water bath

Ester (RCO.OR') (and lactone) or **Anhydride** (RCO.O.COR) — Red to violet coloration

Pink colour fades in test solution

Esters. I.R. 1735 cm^{-1}, R and R' = Alk, ν C=O (s); 1725–1715 cm^{-1}, R = Ar or vinyl, ν C=O (s); 1765–1755 cm^{-1}, R' = Ar or vinyl, ν C=O (s)*; 1735 cm^{-1}, R and R' = Ar, ν C=O (s); 1735 cm^{-1}, δ-lactone ν C=O (s); 1770 cm^{-1}, γ-lactone ν C=O (s); 1300–1050 cm^{-1}, asym. and sym. ν C—O—C (2 bands, s)
*N.B. Phthalate esters have ν C=O at 1780–1760 cm^{-1}
N.M.R. See Table XII

Anhydrides. I.R. 1830–1810 cm^{-1}, 1770–1750 cm^{-1}, R = Alk, ν C=O (doublet due to coupling, s); 1795–1775 cm^{-1}, 1735–1715 cm^{-1}, R = Ar or vinyl, ν C=O (doublet due to coupling, s); 1810–1790 cm^{-1}, 1760–1740 cm^{-1}, ν C=O (6-membered cyclic anhydride, s); 1875–1855 cm^{-1}, 1800–1775 cm^{-1}, ν C=O (5-membered cyclic anhydride)
N.M.R. See comment relating to esters

Table I. Functional Groups Containing C, H, O (*cont.*)

Chemical test	Observation	Functional group	Spectral data
a. If Test 5 is positive, dissolve the compound in benzene or trichloromethane (chloroform) and add aniline (2 drops). Warm gently for 1–2 min	Precipitate formed	**Anhydride**	

Notes

5. (i) Esters and anhydrides react with hydroxylamine to give the corresponding hydroxamic acid:

$$RCO_2R' \xrightarrow{\quad NH_2OH \quad} RC{\overset{\overset{\displaystyle NOH}{\|}}{-}}OH \rightleftharpoons RC{\overset{\overset{\displaystyle O}{\|}}{-}}NHOH$$

The hydroxamic acid gives coloured complexes with iron(III) (cf. enols).

(ii) When potassium hydroxide is consumed in the usual hydrolysis of an ester or anhydride, the pH of the mixture decreases as the reaction proceeds. This is indicated by the colour change of the indicator, phenolphthalein.

a. Aniline reacts readily with anhydrides to give the corresponding anilide, e.g., $PhNH_2 + (CH_3CO)_2O \rightarrow PhNHCOCH_3$. The anilides are only sparingly soluble in benzene and trichloromethane and appear as precipitates.

6. (i) Visual examination	Compound has a red to yellow colour when pure	**Quinone** 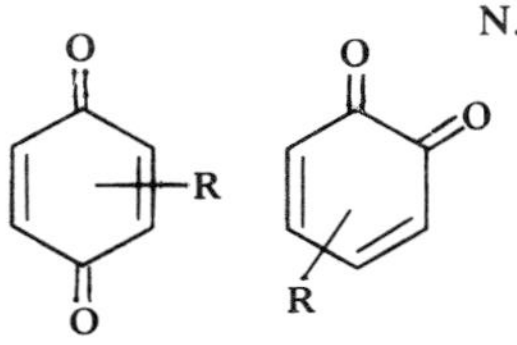	I.R. 1690–1655 cm^{-1}, ν C$=$O N.M.R. δ 6.5–7.5 (protons in quinone ring)
(ii) Add 2 M sodium hydroxide	Pronounced intensification of the original colour		

18

Notes

6. (i) Compounds having conjugated quinonoid structures absorb in the visible range of the spectrum as a result of a weak n → π* transition and are correspondingly coloured.
(ii) Enhancement of the original coloration by a base results from the formation of a carbanion of the type

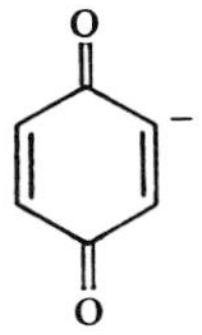

The increased electron availability results in increases in both ε_{max} and λ_{max}.

| 7. Warm at 40–50° with concentrated sulphuric acid | Compound dissolves completely without charring | **Ether** (ROR') | I.R. 1275–1200 cm^{-1}, R = Ar or vinyl asym. ν C—O—C (s); 1150–1070 cm^{-1}, R = Alk asym. ν C—O—C (s)
 N.B. Absorptions in the regions given above are not necessarily conclusive evidence of an ether as such but only of the C—O—C grouping
 N.M.R. See Table XII for effect of ether linkage on chemical shift of nearby protons |

Note

7. Ethers are generally unreactive compounds and will usually dissolve in warm concentrated sulphuric acid without decomposition (charring). Note that the presence of other functional groups may invalidate this test and caution should be used in its interpretation.

Table I. Functional Groups Containing C, H, O (*cont.*)

Chemical test	Functional group	Observation	Spectral data
8. (i) Treat with 1% potassium permanganate solution and shake well N.B. This test is not valid for compounds containing readily oxidized groups.	**Alkene** (RCH=CHR') **Alkyne** (RC≡CR')	Purple colour is discharged rapidly in cold	**Alkene** I.R. 1670–1610 cm^{-1}, ν C=C (variable intensity, absent if compound symmetrical about double bond); 1600–1590 cm^{-1}, ν C=C (conjugated, m). For C—H data, see Table XI N.M.R. δ 4.5–6.3
(ii) Dissolve the compound in carbon tetrachloride and add 5% bromine in carbon tetrachloride (2 drops)		Red-brown colour is discharged *without evolution of hydrogen bromide*	**Alkyne** I.R. 2260–2190 cm^{-1}, ν C≡C (disubstituted, w) 2140–2100 cm^{-1}, ν C≡C (monosubstituted, w). N.M.R. δ 2.5–3.0

Notes

8. (i) The oxidizing agent is reduced by reaction with double and triple bonds with a loss of purple colour:

$$\text{RCH=CHR'} \xrightarrow[\text{H}_2\text{O}]{\text{[O]}} \text{RCHOH . CHOHR'}$$

$$\text{RC≡CR'} \xrightarrow[\text{H}_2\text{O}]{\text{[O]}} \text{RCO.COR'}$$

Note that this test may be positive when other readily oxidized groups are present. In such cases, the bromine test (below) should be carried out.
(ii) Bromine adds readily to alkenes and alkynes with the loss of the red-brown bromine colour:

$$\text{RCH=CHR'} \xrightarrow{\text{Br}_2} \text{RCHBr.CHBrR'}; \quad \text{RC≡CR'} \xrightarrow{\text{Br}_2} \text{RCBr=CBrR'}$$

This test may give ambiguous results with functions containing readily replaceable hydrogen atoms or a reactive aromatic ring when hydrogen bromide is evolved.

9. All of the above tests are negative	Other **hydrocarbons**		I.R. 1650–1450 cm^{-1} arom., ν C=C (up to three peaks present, but best diagnostic peak at ~1600 cm^{-1}); 1250–1200 cm^{-1}, 1170–1145 cm^{-1}, ν C—C (skeletal mode doublet, variable); for C—H data, see Table XI N.M.R. Saturated alkanes δ 0.4–2.1; benzenoid δ 6.0–8.6

Note
9. Aromatic and saturated aliphatic hydrocarbons are unreactive under the conditions of the above tests.

Table II. Functional Groups Containing C, H, N, O

Chemical test	Observation	Functional group	Spectral data
1a. Dissolve* in 2 M hydrochloric acid at room temperature, cool to 5° in ice and add 5% aqueous sodium nitrite (3 or 4 drops) *A few weakly basic amines require concentrated hydrochloric acid; if this fails, the amine should be dissolved in the minimum of ethanol and a little concentrated sulphuric acid should be added and the solution cooled in ice. Compounds which have a pyrrole ring are very weakly basic and do not give the above reactions	Effervescence; nitrogen is evolved and a clear solution is obtained	**Pri.aliphatic amine (AlkNH$_2$), amino acid** ($^-$OOCAlkNH$_3$) or **amide (RCONH$_2$)**	**Amines (primary and secondary, including hydrazines)** I.R. 3500–3000 cm^{-1}, ν N–H (w–m, often a doublet for —NH$_2$ group due to asym. and sym. ν N–H); 1640–1560 cm^{-1}, δ N–H(s); 1360–1250 cm^{-1}, 1280–1180 cm^{-1}, ν C–N (doublet for unsaturated carbon); 1230–1030 cm^{-1}, ν C–N (v)*, 1150–1100 cm^{-1}, ν C–N (as in C—N—C, v); 900–650 cm^{-1}, γ N–H (broad and diffuse) N.B. Tertiary amines do not absorb in the region 3500–3000 cm^{-1} *Doublet for tertiary amines N.M.R. Aliphatic δ 1.5–3.5; benzenoid δ 3.5–5.5; heterocyclic δ 4.5–6.5. Saturated endocyclic NH δ 0.0–1.5. †NH in pyrazole and in imidazole ring, δ 10.0–14.0. Hydrazines show great variation in δ values †NH of a pyrrole ring, δ 7.5–11.0
	No effervescence; clear solution obtained	**Pri.aromatic amine** (ArNH$_2$) or **tertiary amine** (R$_3$N)	
	Dark brown soln. obtained	Tertiary aromatic amine unsubstituted in 4-position	
	No effervescence; cloudy solution or emulsion formed	**Secondary amine** (R$_2$NH) or **arylhydrazine** (ArNHNH$_2$)	
b. If Test 1a gives a clear solution, add a few drops of this to a 5% solution of 2-naphthol dissolved in 2 M sodium hydroxide	Bright red to dark brown precipitate	**Pri.aromatic amine** (ArNH$_2$)	**Amides** I.R. 3300–3050 cm^{-1}, ν N–H (m, often a doublet); 1690–1650 cm^{-1}, ν C=C (Amide I band, s); 1640–1600 cm^{-1}, N–H bending mode (Amide II band); 1420–1405 cm^{-1}, ν C—N (Amide III band)
	No coloration; ignore white to yellow precipitates	**Pri.aliphatic amine** (AlkNH$_2$), tertiary **amine**(R$_3$N), **amino acid**($^-$OOCAlkNH$_3$) or **amide**(RCONH$_2$)	**Amino acids.** I.R. 3100–2600 cm^{-1}, νN–H (m); 1665–1585 cm^{-1}, NH$_3$ deformation (Amino acid I band, w); 1550–1485 cm^{-1}, NH$_3$ deformation (Amino acid II, v). N.M.R. Usually insoluble in CDCl$_3$; spectra taken in D$_2$O which converts NH$_3$ to ND$_3$.
c. Dissolve in 2 M hydrochloric acid at room temperature and add an excess of Fehling's solution. Heat on a boiling-water bath for 5 min	Brick-red precipitate	Substituted **hydrazine**(RNHNH$_2$).	
d. Dissolve in water and add 1% methanolic ninhydrin. Warm gently	Violet coloration	α- or β-**Amino acid**	

Notes

1a. Nitrous acid reacts with amino groups in different ways depending on their nature. Aliphatic amines, amino acids and amides react to give unstable diazonium salts which break down with evolution of nitrogen:

e.g., AlkNH$_2$ $\longrightarrow$ [AlkN$\equiv$NCl$^-$] $\longrightarrow$ AlkOH + N$_2$ + HCl

Table II. Functional Groups Containing C, H, N, O (*cont.*)

Chemical test	Observation	Functional group	Spectral data

The above equation is an oversimplification of the reaction and other organic products are often formed. Aromatic and tertiary amines form diazonium and nitrite salts respectively which are stable below 5°. No nitrogen is evolved unless the solution is warmed above this temperature. Note that tertiary aromatic amines with vacant 4-positions form 4-nitroso compounds which usually give brown solutions:

Secondary amines and hydrazines react with nitrous acid to form *N*-nitroso compounds which appear as yellow oils:

$$R_2NH \longrightarrow R_2N.NO$$

1b. Stable diazonium salts will couple with alkaline 2-naphthol to give brightly coloured (usually red) azo dyes:

1c. Substituted hydrazines are reducing agents and convert copper(II) (Fehling's) solutions to copper(I) oxide (brick-red precipitate).

1d. After elimination of simple primary and secondary amines, amino acids may be identified by their reaction with ninhydrin (1,2,3-indantrione hydrate). The reaction is complex, but may be represented in general terms as follows:

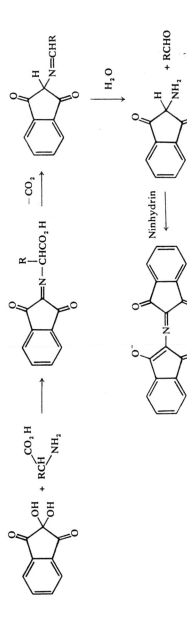

Blue—violet

Table II. Functional Groups Containing C, H, N, O (*cont.*)

Chemical test	Observation	Functional group	Spectral data
2. Add concentrated sodium hydroxide solution. Heat strongly for about 2 min	Ammonia evolved	**Ammonium salt** (RCOONH$_4^+$), **amide** (RCONH$_2$), **imide** (RCONHCOHNR) or **nitrile** (RCN)	**Ammonium salts.** I.R. 3300–3030 cm^{-1}, ν N—H ($\overset{+}{N}$H$_4$ group), 1430–1390 cm^{-1}, bending mode for N—H ($\overset{+}{N}$H$_4$ group) N.M.R. Usually insoluble in CDCl$_3$
a. If Test 2 is positive, add cold 2 M sodium hydroxide solution.	Compound dissolves and ammonia is evolved	Ammonium salt	**Imides.** I.R. Spectra are similar to those of *N*-substituted amides, particularly on low resolution instruments. Higher resolution can show a doublet for ν C=O N.M.R. δ 8.0–10.0, very broad. N.B. The above —NH— or —NH$_2$ groups are converted into —ND— or ND$_2$ by treatment with D$_2$O
b. Mix with sulphur in a dry test tube and heat gently. Test the vapour produced with an absorbent strip treated with 1% aqueous iron(III) nitrate solution	Red stain on paper	Nitrile	**Nitriles.** I.R. 2260–2240 cm^{-1}, R = Alk, ν C≡N (m, very sharp); 2240–2220 cm^{-1}, R = Ar or vinyl, ν C≡N.
c. If Test 2 is positive but Tests 2a and 2b are negative, an imide may be distinguished from an amide by mixing saturated methanolic solutions of the compound and potassium hydroxide	White precipitate formed	Imide	

Notes

2. Ammonium salts, amides, imides and nitriles are hydrolysed by strong alkali with evolution of ammonia:

$$\text{RCN} \xrightarrow[\text{H}_2\text{O}]{\text{NaOH}} [\text{RCONH}_2] \longrightarrow [\text{RCO}_2\text{NH}_4] \longrightarrow \text{RCO}_2\text{Na} + \text{NH}_3$$

nitrile   amide   ammonium salt

a. The ease of hydrolysis decreases as follows: ammonium salts > amides > nitriles; ammonium salts are readily hydrolysed with cold, dilute alkali to give ammonia.

b. Nitriles react with sulphur at high temperature to give thiocyanic acid which can be detected by its reaction with iron(III) giving a red coloration. Note that compounds containing the >C=N— grouping also give this test.

c. Due to the electron attracting influence of the flanking carbonyl groups, the —NH function of an imide is sufficiently acidic to give a methanol-insoluble potassium salt on treatment with potassium hydroxide.

Procedure	Observation	Conclusion	Spectroscopy
3: Add 70% sulphuric acid and reflux for 10 min. Cool in ice to 5° and add 5% sodium nitrite solution. Pour this mixture into a 5% solution of 2-naphthol in 2 M sodium hydroxide.	Bright red to dark brown precipitate	*N*-Substituted **amide** (RCONHAr)	I.R. Similar absorption to those given above for amides, but values for the Amides II and III bands are usually lowered by ~70–150 cm^{-1}; also ν N—H appears as a singlet. Note that if R = Ar or vinyl, values are increased by ~15 cm^{-1} N.M.R. δ 5.0–8.5, very broad for —CONH$_2$ but usually sharp for —CONHR. Two hydrogens of —CONH$_2$ are sometimes not equivalent and give two signals
	No coloured precipitate; ignore white to yellow precipitates	Possibly RCONHAlk	

Notes

3. *N*-Substituted amides are readily hydrolysed in strong acid to give the corresponding amine (as the salt):

$$\text{RCONHR}' \xrightarrow[\text{H}_2\text{O}]{\text{H}^+} \text{RCO}_2\text{H} + \text{R}'\overset{+}{\text{N}}\text{H}_3$$

A primary aromatic amine so formed may then be characterized as described above. (Tests 1a and 1b).

Procedure	Observation	Conclusion	Spectroscopy
4. (i) To an aqueous solution of iron(II) sulphate solution (1 cm^3) add a few drops of 2 M sodium hydroxide followed by the organic compound. Prepare a similar mixture but omit the organic compound. Heat on a boiling-water bath for not more than 2 min.	Grey to green precipitate in test solution changes to brown	**Nitro compound** (RNO$_2$)	**Nitro compounds.** I.R. 1615–1540 cm^{-1}, 1390–1320 cm^{-1}, R = Alk, ν N=O (doublet for asym. and sym. modes, s); 1548–1508 cm^{-1}, 1365–1335 cm^{-1}, R = Ar, ν N=O (s, doublet as above).
(ii) Dissolve the compound in propanone(acetone) and add 5% titanium(III) chloride or sulphate solution (3–5 drops); warm gently	Mauve colour is discharged within 2 min		**Nitrophenols.** I.R. and N.M.R. See individual functional group absorptions.
a. If Test 4 is positive, reduce the compound as follows: add tin and 7 M hydrochloric acid and warm, with continual shaking, for 15 min. Filter the mixture, cool to 5° in ice and add 5% aqueous sodium nitrite. Add this to a 5% solution of 2-naphthol in 2 M sodium hydroxide.	Red to dark brown precipitate	Aromatic **nitro** compound (ArNO$_2$)	
	No coloured precipitate; ignore white to yellow precipitate	Aliphatic **nitro** compound (AlkNO$_2$)	

Table II. Functional Groups Containing C, H, N, O (*cont.*)

Chemical test	Observation	Functional group	Spectral data
b. Add 2 M sodium hydroxide solution	Intense yellow or orange coloration or precipitate	**Nitrophenol** (HOArNO$_2$)	

Notes

4. (i) Iron(II) sulphate reacts with sodium hydroxide to give a grey-green precipitate of iron(II) hydroxide which reduces nitro compounds to amines, and is itself oxidized to brown iron(III) hydroxide.
(ii) Similarly, mauve coloured titanium(III) salts are powerful reducing agents and are oxidized by nitro compounds to colourless titanium(IV).
a. Both aromatic and aliphatic nitro compounds are reduced by tin and hydrochloric acid to the corresponding amines. These may be characterized as described in Tests 1a and 1b.
b. Nitrophenols readily form salts in strong base, the anion acting as a powerful auxochrome in the visible range of the absorption spectrum as a result of the increased electron availability in the ion.

Table III. Functional Group Containing C, H, O, Halogen

Chemical test	Observation	Functional group	Spectral data
1. Prepare a clear saturated solution of the compound in 2 M nitric acid and add aqueous silver nitrate	White precipitate	**Acyl halides** (RCOHal)	I.R. 1810–1780 cm^{-1}, ν C=O (doublet when R = Ar, s)
2. If Test 1 is negative, treat the compound with methanolic potassium hydroxide. Boil for 2 min, cool, acidify with 2 M nitric acid and add 2% ethanolic silver nitrate	White to yellow precipitate formed in cold or on slight warming	**Alkyl halide** (AlkHal)	I.R. 1250–960 cm^{-1}, ν C–F; 830–500 cm^{-1}, ν C–Cl (s, note overtone band at 1510–1480 cm^{-1}); 650–520 cm^{-1}, ν C–Br; 550–400 cm^{-1}, ν C–I
Tests 1 and 2 are negative		**Aryl halide** (ArHal)	

Notes

1, 2. The various types of halide may be distinguished by the ease with which they are hydrolysed. Acyl halides are readily hydrolysed in cold water; alkyl halides require mild alkaline conditions. Most aryl halides are not hydrolysed under these conditions as a result of resonance stabilization of the molecule but the presence of strongly electron-withdrawing functions in *ortho* and *para* positions renders the halogen atom more labile.

Table IV. Functional Groups Containing H, O, S

Chemical test	Observation	Functional group	Spectral data
1. Add water, shake well and test with blue litmus paper	Readily soluble with acid reaction	**Sulphonic acid** (RSO_2OH)	**Sulphonic acids.** I.R. 1250–1160 cm^{-1}, 1080–1000 cm^{-1}, ν S$=$O (doublet due to asym. and sym. stretching modes); 700–610 cm^{-1}, ν S–O. Note that absorption of the O–H group is similar to that in carboxylic acids N.M.R. Usually insoluble in CDCl$_3$; in water, δ 11.0–12.0
2. Odour	Unpleasant and penetrating	**Thiol or thiophenol** (RSH) (also impure **thioether**)	
a. If Test 2 is positive, dissolve in ethanol and add solid sodium nitrite followed by 2 M sulphuric acid	Red coloration	Pri. or secondary **thiol** (AlkSH)	**Thiols and Thiophenols.** I.R. 2600–2500 cm^{-1}, ν S–H (s) N.M.R. Similar to alcohols and phenols (Table I).
	Green coloration changing to red on standing	**Thiophenol** (ArSH)	**Thioethers.** I.R. 695–655 cm^{-1}, 630–600 cm^{-1}, C–S–C (doublet) N.M.R. Similar to ethers
	No coloration	**Thioether** (RSR)	

Notes
1. Sulphonic acids are sufficiently strong acids to turn blue litmus red.
2. Thiols and thiophenols are noted for their particularly unpleasant odour and for this reason are not usually encountered in undergraduate laboratories.

Table V. Functional Groups Containing O, P

Chemical test	Observation	Functional group	Spectral data
1. Reflux with 30% aqueous sodium hydroxide for 20 min and then distill off all the volatile material. Acidify the residue with 2 M sulphuric acid, extract with ether and add a solution of ammonium molybdate in concentrated nitric acid to the aqueous phase. Warm but do not boil	Yellow precipitate obtained	**Phosphate ester** (R_3PO_4)	I.R. 1315–1180 cm^{-1}, ν P$=$O (s); 1195–1185 cm^{-1}, ν P$=$O (w, very sharp if R $=$ CH$_3$); 1100–950 cm^{-1}, ν P–O–C; 950–875 cm^{-1}, (s if R $=$ Ph)

Note
1. Phosphate esters are hydrolysed in strong alkali, liberating phosphate ions. These may be detected (after removal of organic material) by reaction with ammonium molybdate in nitric acid, when a yellow precipitate of ammonium phosphomolybdate is formed.

Table VI. Functional Groups Containing H, Halogen, N

Chemical test	Observation	Functional group	Spectral data
1. Dissolve in water; add excess of nitric acid and then aqueous silver nitrate	White to yellow precipitate formed	**Hydrohalide** salt of a base ($R\overset{+}{N}H_3\bar{X}$, $R_2\overset{+}{N}H_2\bar{X}$, $R_2\overset{+}{N}H\bar{X}$) or **quaternary ammonium salt** ($R_4\overset{+}{N}\bar{X}$)	**Hydrohalides.** I.R. 3000–2250 cm^{-1}, ν N—H (often very broad and complex, particularly for salts of secondary and tertiary amines); 1600–1575 cm^{-1}, N—H bending mode (doublet with band at 1500 cm^{-1} for salts of primary amines). Note that these bands are absent for salts of tertiary amines. N.M.R. Usually insoluble in CDCl$_3$, and should be converted into the free amines (q.v.)
a. If Test 1 is positive, add excess of alkali and extract the mixture with ether. Dry the extract over anhydrous sodium sulphate and evaporate the ether on a water bath	Residue obtained; examine as described in Table II, Tests 1a, 1b and 1c	**Hydrohalide** salt of a base	**Quaternary ammonium salts.** I.R. No characteristic absorptions N.M.R. Usually insoluble in CDCl$_3$
	No residue	**Quaternary ammonium salt**	

Notes
1. Both hydrohalides of bases and quaternary ammonium salts are ionic and dissolve in water liberating halide ions which may be detected using silver nitrate in nitric acid.
a. Treatment of hydrohalides of bases with alkali liberates the organic base which may be characterized (after isolation) as described in Table II, Tests 1a and 1b. Quaternary ammonium salts are not affected by addition of base and are not extractable from the aqueous phase by ether.

Table VII. Functional Group Containing Halogen, O, S

Chemical test	Observation	Functional group	Spectral data
1. Heat with water for 10 min and cool. Acidify with 2 M nitric acid and add aqueous silver nitrate	White to yellow precipitate	**Sulphonyl halide** (RSO$_2$Hal)	I.R. 1385–1320 cm^{-1}, 1185–1150 cm^{-1}, ν S=O (doublet for asym. and sym. modes)

Note
1. Sulphonyl halides are hydrolysed in warm water to give halide ions (cf. alkyl halides).

Table VIII. Functional Groups Containing C, H, N, O, S

Chemical test	Observation	Functional group	Spectral data
1. Heat with solid potassium hydroxide	Ammonia evolved	**Sulphonamide** (RSO_2NH_2)	I.R. 1375–1330 cm^{-1}, 1180–1160 cm^{-1}, S=O (doublet for asym. and sym. modes). Other absorptions are similar to those of amides and N-substituted amides
	Amine evolved	**N-Substituted sulphonamide** (RSO_2NHR' or $RSO_2NR^1R^2$) Examine as described in Table II Tests 1a and 1b	
2. Heat with 2 M sulphuric acid	Hydrogen sulphide evolved	**Thioamide** ($RCSNH_2$)	I.R. 1405–1290 cm^{-1}, C=S (analogous to Amide I band for amides, see Table II)
3. Dissolve in water, acidify with 2 M hydrochloric acid and add barium chloride solution	White precipitate	**Sulphate salt of base** (RNH_3HSO_4, R_2NH_2 HSO_4 or $R_3\overset{+}{N}H\, HSO_4$)	I.R. See data for hydrohalides of bases (Table VI)

Notes

1. Sulphonamides and N-substituted sulphonamides are hydrolysed under basic conditions to give ammonia and various amines respectively. The amines may be characterized as described in Table II, Tests 1a and 1b.

2. Thioamides hydrolyse readily to give hydrogen sulphide and the corresponding amide.

$$\underset{RC.NH_2}{\overset{S}{\|}} \xrightarrow{H_2O, H^+} \underset{RC.NH_2}{\overset{O}{\|}} + H_2S$$

3. Sulphates of bases dissolve in water liberating sulphate ions. These may be detected by addition of barium chloride (in hydrochloric acid), when a white precipitate of barium sulphate is formed.

Table IX. Functional Group Containing H, N, O, P

Chemical test	Observation	Functional group	Spectral data
1. Dissolve in water and treat with concentrated nitric acid and aqueous ammonium molybdate. Warm, but do not boil	Yellow precipitate obtained	Phosphate of a base	I.R. See data for hydrohalides of bases (Table VI)

Note

1. Phosphates of bases dissolve in water liberating phosphate ions (cf. Table V, Note 1).

Table X. Characteristic absorptions of alkane groups

Group	Frequency (cm^{-1})†	Assignment
—CH$_3$	2962 ± 10	Asym. v C—H, s
	2872 ± 10	Sym. v C—H, s
	1450 ± 20	Asym. C—H bending, m
	1380 ± 10	Sym. C—H bending, s
—CH(CH$_3$)$_2$	1385–1370	Doublet, C—H bending, s
	1170 ± 5	v C—C, v
	1145 ± 5	C—C—H Skeletal binding, v
—C(CH$_3$)$_3$	1397–1370	Doublet, C—H bending, m–s
	1250 ± 5 ⎱	C—C Skeletal bending, v
	1210 ± 6 ⎰	
—CH$_2$—	2926 ± 5	Asym. v C—H, s
	2853 ± 5	Sym. v C—H, s
	1465 ± 15	C—H Bending, very sharp, m
	1350–1150	C—H Twisting, v
	1100– 700	C—H Rocking, s
—(CH$_2$)$_n$—	740– 720	$n \geq 4$, C—C bending, singlet in liquid, doublet in solid, v
≧C—H	2890 ± 10	v C—H, w

†Where the range of frequency is given as a ± value, the figure quoted is the most commonly found value. Where a simple range is given, absorptions are widely spread between the two values.

Table XI. Characteristic stretching and bending absorption of alkenic and aryl C—H groups

Group	Frequency (cm^{-1})†	Assignment
=CH$_2$	3080 ± 20	Asym. v C—H, w–m
	2975 ± 10	Sym. v C—H, w–m
Aryl C—H	3050 ± 30	v C—H, w
R'H / C=C / HH (structure)	990 ± 5	γ C—H, m
	907 ± 3	γ C—H, m
R'H / C=C / HR'' (structure)	970 ± 6	γ C—H, m
R'H / C=C / R''H (structure)	890 ± 5	γ C—H, s
R'R''' / C=C / R''H (structure)	815 ± 20	γ C—H, m

$\begin{array}{c} R' \\ \diagdown \\ H \end{array} C{=}C \begin{array}{c} R'' \\ \diagup \\ H \end{array}$	690 ± 20	γ C—H, v, difficult to interpret in some cases
Aryl C—H	671	Benzene, γ C—H, s
Monosubstituted*	900–860	γ C—H, w–m
	770–730	γ C—H, s
	710–690	γ C—H, s
1,2-Disubstituted*	960–905	γ C—H, w
	850–810	γ C—H, w
	760–745	γ C—H, w
1,3-Disubstituted*	960–900	γ C—H, m
	880–830	γ C—H, m–s
	820–790	γ C—H, w–m
1,4- and 1,2,3,4-Substituted*	860–800	γ C—H, s
1,2,3-Trisubstituted*	965–950	γ C—H, w
	900–885	γ C—H, w
	780–760	γ C—H, s
	720–686	γ C—H, m
1,2,4-Trisubstituted*	940–920	γ C—H, w
	900–885	γ C—H, m
	780–760	γ C—H, s
1,3,5-Trisubstituted*	950–925	γ C—H, v
	860–830	γ C—H, s
1,2,3,5-, 1,2,4,5- and 1,2,3,4,5-Polysubstituted*	870–850	γ C—H, s

*These values are most reliable for alkyl substituents.
†Where the range of frequency is given as a ± value, the figure quoted is the most commonly found value. Where a simple range is given, absorptions are widely spread between the two values.

Table XII. Chemical shift of protons attached to carbon (relevant hydrogens in bold type)

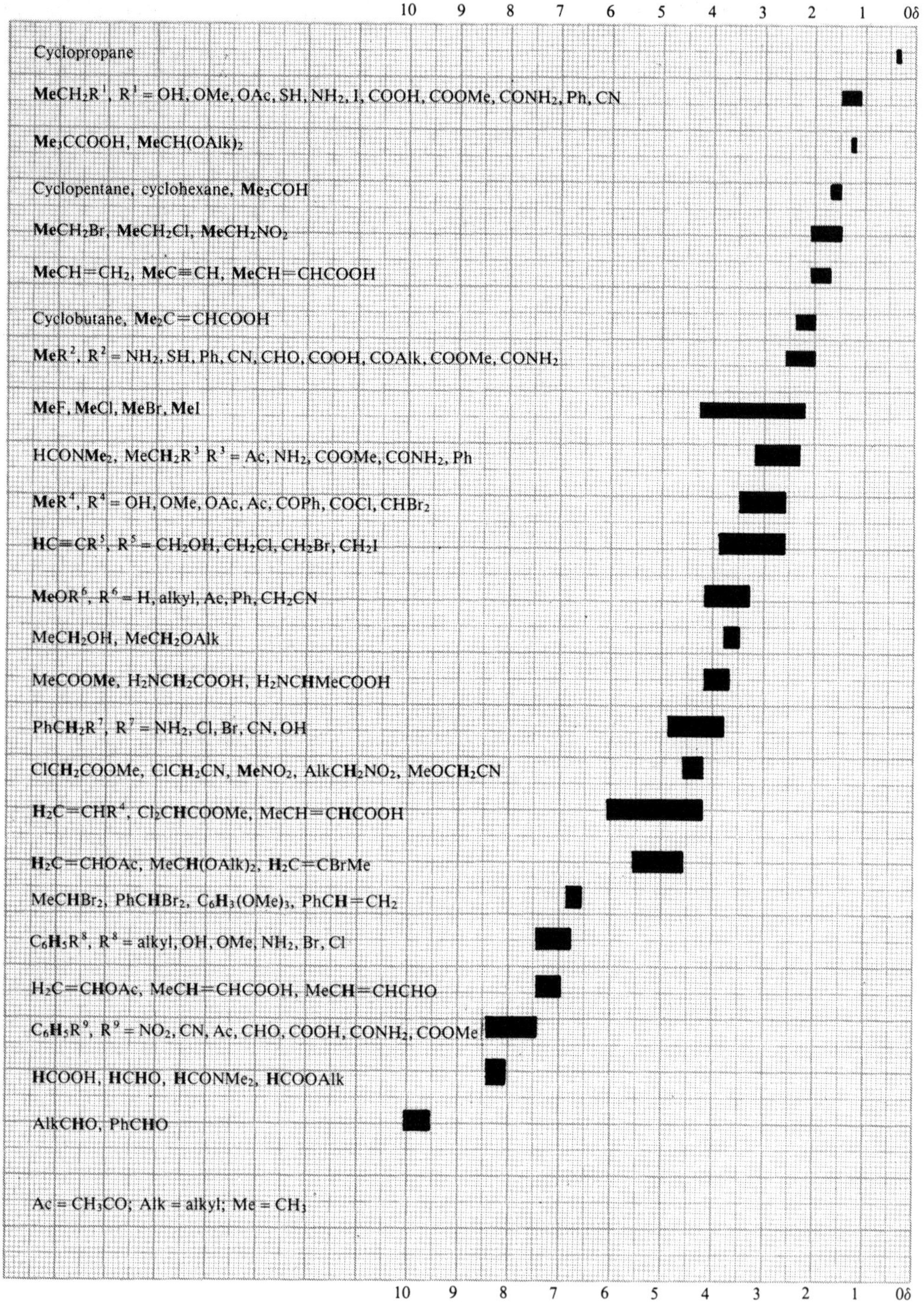

| | 10 | 9 | 8 | 7 | 6 | 5 | 4 | 3 | 2 | 1 | 0δ |

Cyclopropane

MeCH₂R¹, R¹ = OH, OMe, OAc, SH, NH₂, I, COOH, COOMe, CONH₂, Ph, CN

Me₃CCOOH, MeCH(OAlk)₂

Cyclopentane, cyclohexane, Me₃COH

MeCH₂Br, MeCH₂Cl, MeCH₂NO₂

MeCH=CH₂, MeC≡CH, MeCH=CHCOOH

Cyclobutane, Me₂C=CHCOOH

MeR², R² = NH₂, SH, Ph, CN, CHO, COOH, COAlk, COOMe, CONH₂

MeF, MeCl, MeBr, MeI

HCONMe₂, MeCH₂R³ R³ = Ac, NH₂, COOMe, CONH₂, Ph

MeR⁴, R⁴ = OH, OMe, OAc, Ac, COPh, COCl, CHBr₂

HC≡CR⁵, R⁵ = CH₂OH, CH₂Cl, CH₂Br, CH₂I

MeOR⁶, R⁶ = H, alkyl, Ac, Ph, CH₂CN

MeCH₂OH, MeCH₂OAlk

MeCOOMe, H₂NCH₂COOH, H₂NCHMeCOOH

PhCH₂R⁷, R⁷ = NH₂, Cl, Br, CN, OH

ClCH₂COOMe, ClCH₂CN, MeNO₂, AlkCH₂NO₂, MeOCH₂CN

H₂C=CHR⁴, Cl₂CHCOOMe, MeCH=CHCOOH

H₂C=CHOAc, MeCH(OAlk)₂, H₂C=CBrMe

MeCHBr₂, PhCHBr₂, C₆H₃(OMe)₃, PhCH=CH₂

C₆H₅R⁸, R⁸ = alkyl, OH, OMe, NH₂, Br, Cl

H₂C=CHOAc, MeCH=CHCOOH, MeCH=CHCHO

C₆H₅R⁹, R⁹ = NO₂, CN, Ac, CHO, COOH, CONH₂, COOMe

HCOOH, HCHO, HCONMe₂, HCOOAlk

AlkCHO, PhCHO

Ac = CH₃CO; Alk = alkyl; Me = CH₃

| | 10 | 9 | 8 | 7 | 6 | 5 | 4 | 3 | 2 | 1 | 0δ |

32

Table XIII. Chemical shift of protons attached to oxygen, nitrogen or sulphur

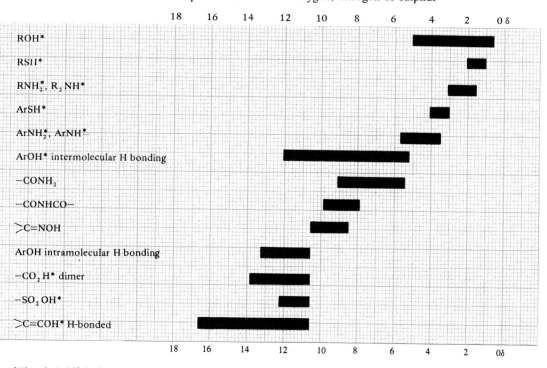

*Chemical shift is dependent on concentration and temperature

Table XIV. Spin–spin coupling constants (Hz)

	Range
$H-\overset{\mid}{\underset{\mid}{C}}-\overset{\mid}{\underset{\mid}{C}}-H$	4–10
$H-\overset{\mid}{\underset{\mid}{C}}-OH$	4–9
$H-\overset{\mid}{\underset{\mid}{C}}-NH-$	4–9
$=C\overset{H}{\underset{H}{\diagdown}}$	0–3
$\overset{H}{\diagdown}C=C\overset{H}{\diagup}$	4–14

33

Table XIV. Spin–spin coupling constants (Hz) (*cont.*)

Structure	J	Value
$\begin{array}{c}\text{H}\\ \diagdown \\ \text{C}=\text{C} \\ \diagup \qquad \diagdown \\ \qquad\qquad \text{H}\end{array}$		12–18
$\text{H}-\text{C}=\text{C}-\text{CH}_3$		1–2
$\text{H}-\overset{\mid}{\underset{\mid}{\text{C}}}-\overset{\parallel}{\text{C}}-\text{H}\;(\text{O})$		0.5–3
$-\text{C}=\overset{\mid}{\text{C}}-\overset{\parallel}{\text{C}}-\text{H}$ (H, O)		6–9
benzene (1,2,3,4)	$J_{1,2}$ $J_{1,3}$ $J_{1,4}$	6–10 0.5–3 0–1
furan (O, 2, 3, 4)	$J_{2,3}$ $J_{3,4}$	1–2 3–4
pyridine (N, 2, 3, 4)	$J_{2,3}$ $J_{3,4}$	5–6 7–9

3

CHROMATOGRAPHIC METHODS OF ANALYSIS AND SEPARATION

Chromatography as a technique is now probably the most widely used separative and analytical method available to chemists. Because of the wide diversity of procedures available, it is not possible to cover all aspects of the method in a text of this size, and only those procedures which would be readily applicable in the undergraduate organic laboratory are considered here. The procedures chosen include analytical thin-layer, high performance liquid, gas-liquid and preparative column chromatography.

Paper chromatography has not been included in this text as it has been largely superseded by thin-layer methods, and because separations are usually rather slow. Electrophoresis has also been excluded as its application is not general, being largely applied to the separation of ionic substances.

THIN LAYER CHROMATOGRAPHY (TLC)

In TLC, a thin layer of a suitable adsorbant (stationary phase) is spread onto a glass (sometimes aluminium or a plastic film) plate. The adsorbant, which is usually alumina or silica gel with particle sizes in the range 5–50 μm, normally contains a binding agent (usually plaster of Paris or polyvinyl alcohol) to promote firm bonding of the adsorbant to the plate or film.

Pre-prepared plates and films are now available quite cheaply but can easily be made using special spreading equipment which is also available commercially. For analytical work, film thickness is usually around 1 mm but can be made thicker for preparative separations.

The TLC procedure for analysis requires that two choices of experimental parameter be made, viz. the nature of the adsorbant and the choice of the developing solvent (mobile phase, eluent). These will depend on the nature of the mixture to be separated and the following factors should be considered in the choice. Alumina is generally used for the separation of weakly polar compounds while silica gel is preferred when separating more polar substances. The eluting power of the mobile phase depends essentially on its polarity, ranging from hexane at the low end to highly polar aqueous systems. Some trial and error is usually necessary to find an appropriate mobile phase and it is recommended that choices be discussed with the laboratory supervisor. Single solvents are to be preferred to mixed solvent systems as solvent separation can sometimes occur on the plates, giving rise to erratic results.

Compounds are usually characterised by their R_f-values, i.e. the ratio of the distance moved by the unknown compound to the distance moved by the solvent front. Note that such values must be $\leqslant 1$. TLC is a quick inexpensive and effective way of screening mixtures of non-volatile organic compounds and it is highly recommended for analytical studies.

Procedure for TLC

A glass tank suitable for TLC plates of up to 15 cm in length is charged with the chosen mobile phase to a depth of ≈ 0.5 cm and the tank atmosphere allowed to saturate with the solvent vapour and to equilibrate thermally. To facilitate this, the tank should be kept in a draught-free environment and ideally should be thermally insulated.

The mixture to be examined should be dissolved in a suitable solvent to give a solution of $\approx 1\%$ in concentration and the solution ($\approx 10\ \mu l$) applied carefully to the plate using a fine glass capillary tube at a point ≈ 1.5 cm from the bottom of the plate. Care must be taken to prevent damage to the plate and the diameter of the spot formed should not exceed 0.5 cm.

A number of spots, including reference materials if required, may be applied at convenient intervals (≈ 1.5 cm) along the bottom of the plate. The plate is then placed in the tank with the bottom of the plate in the solvent and the tank lid replaced immediately. The plate is allowed to develop until the solvent front is near to the top of the plate and is then removed. The solvent front is marked using a scriber and the plate dried in a fume cupboard using a hot air supply, commonly a simple hair-dryer.

If the compounds separated are coloured, their R_f-values may be measured directly. When colourless compounds have been separated, it will be necessary to treat the plate with various reagents (usually sprayed on) which will react with the unknowns to give coloured spots which, in addition to allowing the measurement of R_f-values can sometimes be of assistance in identifying the unknown. A useful general purpose spray is 5% ethanolic sulphuric acid, which on heating will give black spots on a white background for most organic compounds. For details of reagents for specific applications, the student will need to consult the laboratory supervisor or the extensive literature on the subject.

It is worth noting that TLC can be used quantitatively by accurately measuring the amount of mixture applied to the plate, removing the spot from the plate and obtaining the compound, usually by solvent extraction. The compound may then be assayed by an appropriate method.

For compounds which separate readily, a micro method using microscope slides (instead of the large plates) in conjunction with small developing pots, is widely used and represents a useful economy of scale.

HIGH PERFORMANCE LIQUID CHROMATOGRAPHY (HPLC)

HPLC is in essence an extension of the TLC method described above. However, specially designed and highly efficient adsorbants are used which are packed into stainless steel columns and the mobile phase forced, usually under high pressure, through the column by a special pump and flow control system designed to give uniform flow through the column. Separation takes place and the substances separated are fed to a detector as they emerge from the column. The range of detectors for HPLC is continually increasing, but the ones most commonly used are the *differential refractometer*, which detects compounds by measuring the difference in refractive index in the solvent/compound mixture relative to the solvent alone, and the *ultraviolet detector*, which measures changes in ultraviolet absorption of the mobile phase. Refractive index measurements allow detection down to ≈ 1 ppm while ultraviolet detection can go down to as little as ≈ 0.01 ppm. Because of its versatility, the ultraviolet detector is now the most widely used.

Once a HPLC unit is set up (this will normally be done by the laboratory supervisor), samples may be injected into a specially designed injection port with a microlitre syringe, again specially designed for the purpose. The separation of the mixture is then normally displayed on a chart recorder or VDU as a series of peaks which, with suitable calibration, can be used quantitatively. Qualitative analysis is normally achieved by measuring the time taken after injection of the sample to emergence of a peak (retention time) and comparing the value to retention times of known compounds which have been determined under identical running conditions.

It is not felt appropriate here to describe in detail the actual setting up procedures for commercial HPLC units as these will be dependent to a large degree on which of the many instruments nowadays available, is used.

HPLC can be used for preparative separations as both detectors described are non-destructive. However, if very high resolution is required, very small samples must be used and preparative studies under these conditions can be tedious, although auto-injection and collection systems are available.

GAS–LIQUID CHROMATOGRAPHY (GLC)

In many senses, GLC and HPLC are similar, particularly in the way in which samples are manipulated and separation data are displayed. However, there are some major differences which should be noted in the procedures. The columns used in GLC are normally of two types, packed and capillary. The former are usually stainless steel or glass, and up to ≈ 5 m in length. They are packed with an inert support material having a high surface area to mass ratio and which is coated with a non-volatile liquid (stationary phase) which must be inert to the materials being separated. Capillary columns are open tubes, usually 10–150 m long, having an internal diameter of ≈ 0.1 mm and with the stationary phase coated on to the internal wall of the tube.

Packed columns, which one would normally find in the undergraduate laboratory, are generally used for routine separation where the higher separative performance of capillary columns is not required.

The mobile phase is a gas, usually nitrogen, although helium and hydrogen are sometimes used, and is forced through the column under pressure.

Unlike HPLC, the GLC column must be kept at constant temperature as retention times are extremely temperature-dependent. Accordingly, the column is enclosed in an oven with efficient temperature control as is the injection port where the sample is volatilized prior to separation. The sample (≈ 1 μl) is injected using a microlitre syringe via a rubber self-sealing septum into a heated injection port. The volatilized sample is then forced through the column by the carrier gas where the components of the mixture are separated according to the differences in partition coefficients of the individual components with respect to the liquid stationary/gaseous mobile two-phase system.

The individual compounds are fed to a detector and the data displayed and interpreted as for HPLC. Detectors used in GLC are numerous, but two are commonly used. These are the *katharometer*, which measures changes in the thermal conductivity of the carrier gas when containing components from the separation, and the *flame ionization detector*, which burns the sample in an air/hydrogen mixture to produce ions. The ions are collected with a high voltage probe and give rise to a current which is a measure of the component detected. It follows, therefore, that GLC may be used both qualitatively and quantitatively as described for HPLC.

GLC is not nowadays commonly used preparatively, but some commercial equipment is available for circumstances which demand that this procedure be used.

As in the case of HPLC, the undergraduate will normally find the GLC equipment set up ready for use and the manufacturer's instructions regarding operation should be rigorously adhered to.

Note that GLC separations are normally limited to compounds which can be volatilized without decomposition, although high molecular mass (polymeric) material can often be characterized by deliberately decomposing the material to low molecular mass fragments to give a characteristic chromatogram. This technique is known as pyrolysis-gas chromatography.

PREPARATIVE COLUMN CHROMATOGRAPHY (PCC)

When the components of a mixture have been separated by TLC or HPLC it is often necessary to scale up the process to provide sufficient material for other characterization procedures to be applied to the unknown, and R_f values and retention data will not always provide an unequivocal identification. Procedures such as the various forms of spectroscopy usually require at least milligram amounts of material for convenient application, and column chromatography can often provide this facility. It should be noted that very good separations of compounds must be achieved on TLC and HPLC separation, as the scaling up process usually results in some loss of resolution. However, similar adsorbants and eluents may be used, although the particle size of the adsorbant in PCC is usually larger than that used in TLC and HPLC.

Procedure for column chromatography

Packed columns are normally prepared in a glass tube having a PTFE tap at the lower end to control fluid flow, a glass sinter just above the tap and fitted at the top with a Quickfit joint of suitable size to mate with a dropping funnel. Column dimensions will depend on the nature and amount of material being separated and these parameters should be discussed with the laboratory supervisor. The glass tube must be thoroughly cleaned prior to use, as traces of grease can cause irregular flow in the column with resulting loss of resolution.

When suitable parameters have been established, the absorbant is treated with sufficient eluent to form a creamy slurry which should then be poured quickly into the column with the tap open. After about 5 seconds, the tap should be shut and the adsorbant allowed to settle. Tapping the tube with a plastic rod will assist in eliminating bubbles in the column and hence allow even settling of the adsorbant.

The solvent may then be run from the column leaving a depth of $\approx 1-2$ cm of solvent above the packing and *it is essential that the solvent never drops below the top of the adsorbant* as this will allow air to enter the column and destroy the evenness of the packing. The tap is turned off and a small cotton, quartz or glass wool plug is then placed carefully on top of the adsorbant. The sample, which should be dissolved in the minimum volume of solvent, in added to the column, the plug preventing disturbance of the column packing. The column tap is opened and the sample allowed to run onto the packing material to almost plug depth and the tap turned off. The column is then fitted with a solvent reservoir, usually a Quickfit dropping funnel containing solvent. The dropping funnel tap and column tap are opened and, when correctly set up, the solvent level in the column will remain constant. Additional solvent, or different solvents, should be added to the reservoir.

The solvent is collected as it emerges from the column and can be collected, usually in standard aliquots, for further examination. Those fractions found to contain separated material are evaporated to dryness on a vacuum rotary evaporator and the residues characterised by appropriate methods.

4

THE SEPARATION OF ORGANIC MIXTURES

A mixture of organic compounds may be in the solid or liquid form or may consist of a solid dissolved or suspended in a liquid. If a solid and a liquid are present it is usually unwise to expect separation to be accomplished by filtration because the liquid phase almost certainly contains some dissolved solid and traces of the liquid component may be difficult to remove from the solid compound. The methods of isolating pure samples of the components from a mixture may be either physical or chemical. The physical method consists of fractional distillation and is applicable only if there is a wide difference between the boiling points of the two compounds and provided that an azeotrope is not formed. The chemical method of separating two compounds depends on their differing solubility in water, ether, dilute acid or alkali.

The procedures described below should be followed in the order given and should be successful for the great majority of mixtures.

1. If the mixture is a liquid, it should be placed in a small flask equipped with a stillhead, thermometer and condenser. Heat the flask carefully; observe whether a liquid distils, and if it does, note the temperature and continue the distillation until the temperature at the stillhead falls, indicating that all the liquid at that temperature has distilled over. The lower-boiling component of the mixture will now be in the receiver flask.

2. If the mixture is a liquid which cannot be separated according to paragraph 1, or if the mixture is a solid, test its solubility in ether. Most organic compounds are soluble but amongst those which have a low solubility are the following: carbohydrates, amino-acids, sulphonic acids, salts of amines, metal salts of carboxylic and sulphonic acids, some aromatic polybasic acids, some amides and ureas, and polyhydroxy compounds.

Liquid mixtures: proceed to test (b) below.

Solid mixtures: if the mixture dissolves completely, proceed to test (b) below. If an insoluble part remains, continue as described in (a). If both components are insoluble, use the sequence of tests given in paragraph 5 below.

(a) Filter off the undissolved solid, keep the ethereal filtrate (A). Allow the solid collected to dry off in air or over slight heat.

Take the filtrate (A) and if all the mixture did not dissolve in ether initially, evaporate the ether over a hot water bath. If a residue (solid or liquid) is obtained, a separation has been achieved by virtue of the ether-solubility of one of the two compounds.

(b) If all the mixture dissolved in ether, place the solution in a tap funnel. Extract this with 10% sodium hydroxide solution, separate the basic extract from the ether layer (B) into a small flask and acidify with 10% hydrochloric acid solution. The appearance of a solid or an oil or of turbidity may indicate the presence of a carboxylic acid or a phenolic compound.

Take the ether layer B and extract it with 10% hydrochloric acid solution. Separate the acid layer (keep the ethereal solution (C)) and basify it with dilute sodium hydroxide. If an oil or a

39

solid separates, extract this twice with ether. Dry the extract with anhydrous sodium sulphate, allow to stand for 10 min in a stoppered flask, filter, evaporate the ether over a hot-water bath and the residue, if any, will be the basic component.

The ethereal layer (C) contains a neutral compound if present in the original mixture; it should be dried with sodium sulphate, filtered and the solvent distilled. The most common classes of neutral compounds are: hydrocarbons, ethers, halides, alcohols, aldehydes, ketones, amides, nitriles, esters, unreactive anhydrides and nitro compounds.

If basic and neutral components are absent, a carboxylic acid and a phenol may be separated as follows: to this mixture add an excess of solid sodium bicarbonate in small amounts with constant stirring until the solution is no longer acid to litmus. Extract the solution with ether; the ethereal layer will contain a phenolic compound if present while the aqueous layer will contain a carboxylic acid.* Acidify this aqueous solution with dilute hydrochloric acid. A solid carboxylic acid may separate out and should be filtered off and dried. If a liquid acid is present, some solid calcium chloride should be added with thorough shaking until the water is almost saturated with it. The acid is then extracted with ether and isolated in the usual way.

3. The procedure given in paragraph 2 above will not separate two neutral compounds. If nothing is extracted from the ethereal solution by dilute acid and dilute alkali, the presence of two neutral compounds should be suspected. If one of these is a carbonyl compound which forms a bisulphite adduct, this may be removed as its addition product. Prepare a 40% solution of sodium metabisulphite in water and add one fifth of its volume of ethanol. If some of the salt separates, filter the solution and to the filtrate (12 cm^3) add the mixture (4 g). Shake thoroughly and allow to cool. The crystalline adduct if formed should be filtered off (keep the filtrate (D)) and decomposed by careful distillation with an excess of sodium carbonate solution. The carbonyl compound if volatile is found in the distillate; otherwise it will have to be extracted out of the mixture in the distilling flask with ether and isolated in the usual way.

The other neutral component will be found in the filtrate D and should be isolated by removing any ethanol that may remain and then extracted with ether.

4. A few compounds which are very soluble in water and in ether can be troublesome to isolate from mixture. For example, 1,3-dihydroxybenzene mixed with an amine or a carbonyl compound would not be readily separated by the procedure given in paragraph 2. 1,3-Dihydroxybenzene is about equally soluble in water as in ether and although it would be extracted out of ether with several portions of dilute sodium hydroxide solution, acidification of this extract will not result in separation of the compound because of its solubility in water. A mixture consisting of an amine and resorcinol may be separated as follows.

Dissolve the mixture in an excess of dilute hydrochloric acid; test the solution with litmus paper. Place the solution in a tap funnel and extract it with three portions of ether each equal to about one half the volume of the mixture. Retain the acidic solution (E). Dry the ethereal extract with anhydrous sodium sulphate, filter and distil off the solvent. The residue should be tested for a phenolic group. Basify the acidic solution E with sodium hydroxide solution, extract with ether. Dry the extract and distil off the solvent. A residue indicates the presence of an amine.

A water-soluble phenol or polyhydric phenol and a carbonyl compound are best separated by converting the latter into its bisulphite compound (see paragraph 3) or its semicarbazone and

---

*Some nitro- and halogeno-phenols are sufficiently acidic to release carbon dioxide from sodium bicarbonate. Therefore an aromatic compound which contains nitrogen or a halogen and which behaves as carboxylic acid in this separation should be tested for phenolic properties.

regenerating it by heating with twice its weight of oxalic acid and ten times its weight of water in a distilling flask. The carbonyl compound, if low boiling, may be distilled and collected, or if high boiling, may be extracted with ether from the residue in the distilling flask and characterized in the usual way.

5. The procedures given above will not be effective if both components are insoluble in ether. In such cases the following suggestions should be useful:

(a) Test the mixture to see if one component is soluble in water. If this is so, filter off the insoluble compound and evaporate the filtrate to dryness on a small Bunsen flame. Overheating may cause decomposition or charring. If a carbohydrate is present a syrup will often be obtained which may be difficult to crystallize but tests should be made on the syrup to determine its nature.

(b) Should both components be insoluble in water or both soluble in water, repeat procedure 5(a) using methanol.

(c) If the above tests have failed to effect a separation, the presence of any two of the following should be suspected: polyhydric alcohol, carbohydrate, metallic salt of a carboxylic acid or a salt of an organic base. Dissolve the mixture in 10% hydrochloric acid and if a solid precipitates, collect on a filter, wash with water and dry carefully. This may be a free acid, probably aromatic. Formation of an oil will indicate a higher aliphatic acid. The aqueous layer may contain a soluble polyol or carbohydrate.

If no solid or oil is obtained, extract the solution several times with ether and evaporate the extract on a water bath. A residue may be a lower aliphatic acid. If nothing was extracted by the ether, make the solution basic with 10% sodium hydroxide; a salt of an organic base if present will lead to the formation of the free base which should be extracted with ether and isolated in the usual way. The aqueous layer may contain a polyol or a carbohydrate together with sodium chloride or sulphate. Removal of the inorganic ions by passing the solution through a mixed-bed ion-exchange column should give an aqueous solution of the polyol or carbohydrate.

PREPARATION OF DERIVATIVES

The identity of an organic compound may be confirmed by converting it into a crystalline derivative of characteristic melting point. The derivative should preferably have a melting point not lower than about 100° because such compounds are usually more readily crystallized and dried.

Purification of derivatives

When preparing a derivative, it is necessary that the product be obtained in a pure form, thus ensuring an accurate melting point. The procedure for recrystallization is as follows: the crude material is dissolved in the minimum of hot solvent, filtered if necessary without suction, and the solution obtained allowed to cool slowly. The crystalline product is collected in a Buchner funnel, washed with a small volume of ice-cold solvent, dried thoroughly and the melting point determined. This procedure should be repeated until no further increase in melting point is observed.

If the crude material is badly discoloured, it is often advantageous to add a little activated charcoal to the solution which should then be boiled gently for up to 5 min. The hot solution is filtered under slight suction and the filtrate allowed to cool slowly.

The choice of a suitable solvent for recrystallization is of considerable importance in the purification of derivatives. Ideally, the compound should have a high solubility in the hot solvent and a low solubility in the cold solvent. If two or more solvents meet this requirement, that with the lowest boiling point should be chosen to facilitate removal from the solid product. It is sometimes impossible to find a satisfactory solvent and then the technique of mixed solvents is often of value. The derivative is dissolved in a slight excess of hot solvent and a second solvent which is miscible with the first solvent but in which the derivative is insoluble, is carefully added with shaking and warming until a faint cloudiness persists. This should disappear on boiling and the solution is allowed to cool slowly.

Experimental details are given below for preparing the derivatives listed in Chapter 6 for the various classes of organic compounds.

ACETALS

(a) Hydrolysis to aldehyde and alcohol

$$RCH(OAlk)_2 \xrightarrow{\text{aq. acid}} RCHO + 2\ AlkOH$$

$$AlkOH + CS_2 + NaOH \longrightarrow AlkOCS_2^-Na^+$$

Treat the acetal (0.5 g) with M hydrochloric acid (5 cm^3) and reflux for up to 30 min. If the resulting solution is homogeneous, divide into two equal portions and characterize the aldehyde as the 2,4-dinitrophenylhydrazone (see Aldehydes, p. 44) and the alcohol as the potassium alkyl xanthate as follows. To a portion from the hydrolysis, add solid potassium hydroxide (7 g) and cool to 40°. Transfer to a separating funnel and add carbon disulphide (3 cm³) (INFLAMMABLE) and propanone (acetone) (3 cm³). Mix cautiously and then shake vigorously for 15 min. Allow to settle, discard the lower layer and filter the remaining solution through glass wool. Precipitate the xanthate with ether and recrystallize from ethanol.

If two layers separate, they should be treated individually as described above.

ALCOHOLS

(a) Ethanoate (acetate)

$$AlkOH + (CH_3 \cdot CO)_2O \longrightarrow AlkO \cdot CO \cdot CH_3 + CH_3 \cdot CO_2H$$

To the compound (0.5 g) add anhydrous sodium ethanoate (acetate) (0.5 g) and ethanoic (acetic) anhydride (3 cm³). Reflux for 20 min and pour into water (25 cm³). Stir until a solid is obtained, filter this off and wash well with water. Recrystallize from ethanol.

(b) Benzoate and toluene-4-sulphonate (Schotten–Baumann method)

$$AlkOH + C_6H_5 \cdot COCl \longrightarrow AlkO \cdot CO \cdot C_6H_5 + HCl$$

$$AlkOH + 4\text{-}CH_3 \cdot C_6H_4 \cdot SO_2Cl \longrightarrow AlkO \cdot SO_2 \cdot C_6H_4 \cdot 4\text{-}CH_3 + HCl$$

Treat the compound (0.5 g) with 2 M sodium hydroxide (10 cm³) and add benzoyl chloride (1 cm³) or toluene-4-sulphonyl chloride (1 g). In the latter case, add just sufficient propanone (acetone) to render the mixture homogeneous. Shake vigorously in a stoppered tube until a solid is obtained. If propanone has been used, it may be necessary to add water in order to precipitate the product. Filter this off, wash well with water and recrystallize from ethanol.

(c) Benzoate and toluene-4-sulphonate (alternative procedure); 4-nitrobenzoate and 3,5-dinitrobenzoate

$$AlkOH + ArCOCl \longrightarrow AlkO \cdot COAr + HCl$$

where Ar = C_6H_5-, $4\text{-}CH_3 \cdot C_6H_4SO_2-$, $4\text{-}NO_2 \cdot C_6H_4-$ or $3,5\text{-}(NO_2)_2 \cdot C_6H_3-$.

Dissolve the compound (0.5 g) in dry pyridine (5 cm³) and add 4-nitrobenzoyl chloride (1 g) or 3,5-dinitrobenzoyl chloride (1.3 g). Reflux for 30 min and pour into 2 M hydrochloric acid (40 cm³). Separate the solid (sometimes an oil is formed) and stir with M sodium carbonate solution (10 cm³). Filter off the solid obtained and recrystallize from ethanol, aqueous ethanol or light petroleum (60–80°).

(d) Alkyl hydrogen 3-nitrophthalate

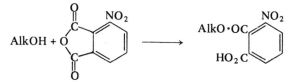

Heat a mixture of the compound (0.5 g) and 3-nitrophthalic anhydride (0.5 g) until a liquid is obtained. Continue heating for a further 15 min, cool, and recrystallize the solid obtained from water or aqueous ethanol. If the original alcohol has a boiling point in excess of 150°, it is advisable to dissolve the mixture in toluene (2–5 cm³) and reflux for up to 30 min. If no solid product is obtained on cooling, precipitate the product by addition of light petroleum (60–80°).

(e) Oxidation of primary alcohols* to carboxylic acid

$$RCH_2 \cdot OH \xrightarrow{[O]} RCO_2H$$

Treat the compound (1 g) with a chromic acid oxidation mixture (10 cm³) consisting of 50% sodium dichromate solution in 6 M sulphuric acid. Reflux until the colour has changed from red to green and add more oxidizing mixture. Continue this procedure until no further discharge of the red colour is observed. Cool the solution, filter off the resulting product and wash well with M sulphuric acid and then with water. Dissolve the product in sodium carbonate solution, filter, and acidify with M sulphuric acid. Filter off the solid, wash well with water and recrystallize from water or ethanol.

ALDEHYDES

(a) 2,4-Dinitrophenylhydrazone

$$RCHO + 2,4\text{-}(NO_2)_2C_6H_3 \cdot NH \cdot NH_2 \longrightarrow RCH{:}N \cdot NH \cdot C_6H_3\text{-}2,4\text{-}(NO_2)_2 + H_2O$$

To the compound (0.5 g) dissolved in ethanol (0.5 cm³) add a solution (2 cm³) of 2,4-dinitrophenylhydrazine (see below) and boil for 2 min. Filter off the resulting precipitate, wash with a little cold ethanol and recrystallize from ethanol, ethanoic (acetic) acid, ethyl ethanoate (acetate) or trichloromethane (chloroform).

Preparation of reagent

(i) Prepare a saturated solution in either 5 M hydrochloric acid or 3 M sulphuric acid. (ii) Dissolve the 2,4-dinitrophenylhydrazine (2 g) in methanol (30 cm³) and water (10 cm³). Add concentrated sulphuric acid (4 cm³) cautiously with shaking. Cool and filter if necessary. (iii) Dissolve 2,4-dinitrophenylhydrazine in a solution of 85% phosphoric acid (60 cm³) in ethanol (40 cm³), heating gently if necessary.

(b) Semicarbazone

$$RCHO + H_2N \cdot NH \cdot CO \cdot NH_2 \longrightarrow RCH{:}N \cdot NH \cdot CO \cdot NH_2 + H_2O$$

To a solution of the compound (0.5 g) in water (2 cm³) add hydrated sodium ethanoate (acetate) (0.75 g) and semicarbazide hydrochloride (0.5 g). Add ethanol dropwise if the solution is not completely clear but care should be taken to add the minimum amount of ethanol otherwise sodium chloride may be precipitated. Heat for up to 10 min on a boiling-water bath, cool and filter. Recrystallize the product from ethanol, water, benzene or ethanoic (acetic) acid.

---

*This method may also be used to oxidize other compounds. e.g.. alkylbenzenes: $ArAlk \xrightarrow{[O]} ArCO_2H$

(c) Oxime

$$RCHO + H_2N \cdot OH \longrightarrow RCH:N \cdot OH + H_2O$$

Dissolve hydroxylamine hydrochloride (0.5 g) in water (3 cm³) and add hydrated sodium ethanoate (acetate) (0.5 g) followed by the aldehyde (0.5 g). Heat on a boiling-water bath and add ethanol dropwise, if necessary, to clear the solution. Continue heating for 1–2 hours and allow to cool. Filter off the solid and recrystallize from ethanol.

(d) 4-Nitrophenylhydrazone (and phenylhydrazone)

$$RCHO + 4\text{-}NO_2 \cdot C_6H_4 \cdot NH \cdot NH_2 \longrightarrow RCH:N \cdot NH \cdot C_6H_4 \cdot 4\text{-}NO_2 + H_2O$$

Prepare a solution of 4-nitrophenylhydrazine (0.5 g) in ethanol (15 cm³) and ethanoic (acetic) acid (0.5 cm³) and add the organic compound (0.5 g) to the solution. Reflux for 10 min, cool and recrystallize the solid product from ethanol. Occasionally, no solid is obtained on cooling. In such cases, the solution should be reheated and water added until a faint cloudiness is observed. Recrystallize the product using this technique (mixed solvents).

For the preparation of phenylhydrazones, replace the 4-nitrophenylhydrazine with phenyl-hydrazine (0.5 g).

(e) Dimethone

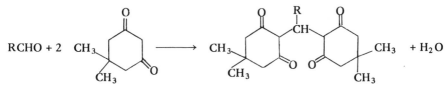

Treat the aldehyde (0.5 g) with a 5% solution of 5,5-dimethylcyclohexane-1,3-dione (dimedone) in 50:50 aqueous ethanol (10 cm³). If no precipitate forms within 2 min, warm the solution for 5 min, cool in ice and filter off the product. Recrystallize from aqueous ethanol or ethanol.

AMIDES, IMIDES, UREAS AND GUANIDINES

(a) Xanthyl derivative

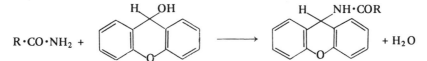

To the organic compound (0.5 g) add a 7% solution of xanthydrol in ethanoic (acetic) acid (7 cm³) and reflux for up to 30 min. Add water (5 cm³) and allow to cool. Filter off the solid product and recrystallize from aqueous dioxan or ethanoic acid.

(b) Diphenylmethyl derivative

$$RSO_2NH_2 + Ph_2CHOH \longrightarrow RSO_2NHCHPh_2 + H_2O$$

Heat a mixture of the sample (0.5 g), diphenylmethanol (0.5 g), toluene-4-sulphonic acid (0.5 g) and acetic acid (5 cm³) under reflux for 30 min. Pour the mixture into water (about 50 cm³), and collect the precipitated solid. This may be recrystallized if necessary from ethanol or toluene.

(c) Hydrolysis to carboxylic acid

$$RCO \cdot NH_2 + NaOH \longrightarrow RCO_2Na + NH_3 \xrightarrow{\text{aq. acid}} RCO_2H$$

$$R\underset{CO}{\overset{CO}{\diagdown}}NH + 2\,NaOH \longrightarrow R\underset{CO_2Na}{\overset{CO_2Na}{\diagdown}} + NH_3 \xrightarrow{\text{aq. acid}} R(CO_2H)_2$$

Reflux the organic compound (0.5 g) with an excess of 6 M sodium hydroxide solution until no further evolution of ammonia is detectable. Acidify the resulting solution with concentrated hydrochloric acid and filter off the solid product obtained. Wash well with water and recrystallize from water, aqueous ethanol or ethanol.

AMIDES, N-SUBSTITUTED

(a) Hydrolysis to carboxylic acid and amine'

$$RCO \cdot NHR' \xrightarrow{\text{aq. acid}} RCO_2H + R'NH_2$$

Treat the organic compound (1 g) with 12 M sulphuric acid (4 cm³) and reflux for 30 min. Cool, basify with sodium hydroxide solution and extract the liberated base with ether. Obtain the free base from the ethereal extract by evaporating the ether on a boiling-water bath and characterize as described under *Amines, Primary and Secondary* (see below). Acidify the remaining aqueous solution with hydrochloric acid and filter off any solid produced. Recrystallize from water, aqueous ethanol or ethanol to obtain the pure carboxylic acid and characterize as described under *Carboxylic Acids* (p. 49). If no solid separates, saturate the solution with sodium chloride and extract with ether. Evaporate the ether and characterize the residue as described under *Carboxylic Acids* (p. 49).

AMINES, PRIMARY AND SECONDARY

(a) Ethanoyl (acetyl) derivative

$$RNH_2 + (CH_3 \cdot CO)_2O \longrightarrow RNH \cdot CO \cdot CH_3 + CH_3 \cdot CO_2H$$

$$RR'NH + (CH_3 \cdot CO)_2O \longrightarrow RR'N \cdot CO \cdot CH_3 + CH_3 \cdot CO_2H$$

Suspend the amine (0.5 g) in water (0.5 cm³) and add a mixture of ethanoic (acetic) acid (0.5 cm³) and ethanoic (acetic) anhydride (0.5 cm³). Heat gently if reaction does not occur immediately. Cool and filter off any solid which separates. If no solid is obtained, neutralize the solution with saturated sodium carbonate solution. Filter off the solid obtained. In either case, recrystallize the product from water or aqueous ethanol. To ethanoylate (acetylate) weakly basic amines, replace the water by a few drops of concentrated sulphuric acid and reflux for 20 min.

(b) Benzoyl, benzenesulphonyl and toluene-4-sulphonyl derivatives

$$RNH_2 + C_6H_5 \cdot COCl \longrightarrow RNH \cdot CO \cdot C_6H_5 + HCl$$

$$RR'NH + C_6H_5 \cdot COCl \longrightarrow RR'N \cdot CO \cdot C_6H_5 + HCl$$

$$RNH_2 + ArSO_2Cl \longrightarrow RNH \cdot SO_2Ar + HCl$$

$$RR'NH + ArSO_2Cl \longrightarrow RR'N \cdot SO_2Ar + HCl$$

where $Ar = C_6H_5-$, $4\text{-}CH_3C_6H_4-$.

Prepare as described under *Alcohols* (p. 43).

(c) 2,4-Dinitrophenyl derivative

$$RNH_2 + 2,4\text{-}(NO_2)_2C_6H_3Cl \longrightarrow 2,4\text{-}(NO_2)_2C_6H_3 \cdot NHR + HCl$$

$$RR'NH + 2,4\text{-}(NO_2)_2C_6H_3Cl \longrightarrow 2,4\text{-}(NO_2)_2C_6H_3 \cdot NRR' + HCl$$

Treat the amine (0.5 g) with an equimolar proportion of 1-chloro-2,4-dinitrobenzene (CAUTION: skin irritant) and anhydrous sodium ethanoate (acetate) (1 g); heat on a boiling-water bath for up to 30 min. Cool, and add cold ethanol (3–4 cm³). Filter off the solid obtained and recrystallize from ethanol.

(d) Picrate

$$RR'R''N + 2,4,6\text{-}(NO_2)_3C_6H_2 \cdot OH \longrightarrow [RR'R''NH]^+ [2,4,6\text{-}(NO_2)_3C_6H_2O]^-$$

where R' and/or R'' may be a hydrogen atom.

Dissolve the compound (0.5 g) in ethanol (2 cm³) and treat with a saturated ethanolic solution of picric acid (3 cm³). Warm gently for 1 min and allow to cool. Recrystallize the product from ethanol.

AMINES, TERTIARY

(a) Methiodide

$$RR'R''N + CH_3I \longrightarrow [RR'R''N \cdot CH_3]^+ [I]^-$$

Add methyl iodide (0.5 cm³) to the dry amine (0.5 g) at room temperature and allow to stand for 5 min. Reflux on a boiling-water bath for a further 5 min and then cool in ice. Filter off the solid product (scratch with a glass rod if no solid is obtained) and recrystallize from ethanol or propanone (acetone).

(b) Picrate

$$RR'R''N + 2,4,6\text{-}(NO_2)_3C_6H_2 \cdot OH \longrightarrow [RR'R''NH]^+ [2,4,6\text{-}(NO_2)_3C_6H_2O]^-$$

Prepare as described under *Amines, Primary and Secondary* (above).

(c) 4-Nitroso derivative (for dialkyl tertiary amines with vacant 4-position in aryl group)

$$R_2N \cdot C_6H_5 + HONO \longrightarrow 4\text{-}R_2N \cdot C_6H_4 \cdot NO + H_2O$$

Dissolve the amine (0.5 cm³) in 2 M hydrochloric acid (4 cm³) and cool in ice to 5°. Add dropwise 20% sodium nitrite solution (2 cm³) and allow to stand in the cold for 5 min. Basify with 2 M sodium hydroxide and extract with trichloromethane (chloroform). Dry the extract over anhydrous sodium sulphate and precipitate the derivative by addition of carbon tetrachloride. Filter off the product and recrystallize from ether (INFLAMMABLE).

(d) Methyl toluene-4-sulphonate salt

$$RR'R''N + 4\text{-}CH_3 \cdot C_6H_4 \cdot SO_2 \cdot O \cdot CH_3 \longrightarrow [RR'R''N \cdot CH_3]^+ [4\text{-}CH_3 \cdot C_6H_4 \cdot SO_2 \cdot O]^-$$

To the amine (0.2 cm³) add methyl toluene-4-sulphonate (0.3 g) and benzene (1 cm³) or diethyl ether (1 cm³). Reflux on a water bath for 20 min and cool. Decant the ether (crystals should remain) and add methanol (1 cm³) and ethyl ethanoate (acetate) (5 cm³) for recrystallization.

AMINO-ACIDS

(a) Benzoyl, 3,5-dinitrobenzoyl and toluene-4-sulphonyl derivatives

$$\underset{R \cdot NH_2}{\overset{CO_2H}{|}} + ArCOCl \longrightarrow \underset{R \cdot NH \cdot COAr}{\overset{CO_2H}{|}} + HCl$$

where Ar = C_6H_5- or $3,5\text{-}(NO_2)_2C_6H_3-$.

$$\underset{R \cdot NH_2}{\overset{CO_2H}{|}} + 4\text{-}CH_3 \cdot C_6H_4 \cdot SO_2Cl \longrightarrow \underset{R \cdot NH \cdot SO_2 \cdot C_6H_4 \cdot 4\text{-}CH_3}{\overset{CO_2H}{|}}$$

Prepare as described under *Amines, Primary and Secondary* (using 3,5-dinitrobenzoyl chloride (1 g) to prepare the 3,5-dinitrobenzoyl derivative). In each case acidify the solution with 2 M hydrochloric acid when the reaction is complete. Filter off the solid obtained and recrystallize from water, aqueous ethanol or ethanol.

(b) Picrate

$$\underset{R \cdot NH_2}{\overset{CO_2H}{|}} + 2,4,6\text{-}(NO_2)_3C_6H_2 \cdot OH \longrightarrow \left[\underset{R \cdot NH_3}{\overset{CO_2H}{|}}\right]^+ [2,4,6\text{-}(NO_2)_3C_6H_2O]^-$$

Prepare as described under *Amines, Primary and Secondary* (p. 46).

(c) Ethanoyl (acetyl) derivative

$$\underset{R \cdot NH_2}{\overset{CO_2H}{|}} + (CH_3 \cdot CO)_2O \longrightarrow \underset{R \cdot NH \cdot CO \cdot CH_3}{\overset{CO_2H}{|}} + CH_3 \cdot CO_2H$$

Prepare as described under *Amines, Primary and Secondary* (p. 46).

CARBOHYDRATES

(a) β-Ethanoate (β-acetate), e.g. of D-glucose:

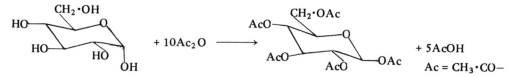

Treat the carbohydrate (0.5 g) with anhydrous sodium ethanoate (acetate) (0.5 g) and ethanoic (acetic) anhydride (3 cm³). Heat on a boiling-water bath for 90 min and pour the product into water (25 cm³). Filter off the solid obtained after stirring, wash well with water and recrystallize from ethanol. If an oil is obtained, decant the water and induce crystallization by scratching with a glass rod.

48

(b) Benzoate (of glucose and fructose only), e.g. of D-glucose:

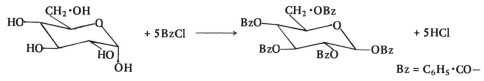

$$Bz = C_6H_5 \cdot CO-$$

Prepare as described under *Alcohols* (p. 43)

(c) 4-*N*-Glycosylaminobenzoic acid, e.g. of D-glucose:

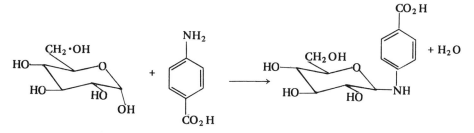

Heat the carbohydrate (1 g) with water (not more than 0.5 cm³) on a boiling-water bath. When most of the sugar has dissolved, add 4-aminobenzoic acid (1 g) in three portions. Continue heating for not more than 2 min. Remove the reaction mixture from the water bath and add methanol (4 cm³). Cool in ice if necessary and filter off the solid. Wash it with a small volume of cold methanol and dry at room temperature in air or under vacuum. The product may be recrystallized from ethanol if desired. The melting point of this derivative is best determined by introducing the sample into the apparatus which has been pre-heated to 120°, or to 170° if an approximate determination of melting point indicated that the compound melted above 180°.

(d) Osazone

$$\begin{array}{l}\text{CHO} \\ | \\ \text{CH}\cdot\text{OH} \\ | \\ \text{R}\end{array} + 2\,C_6H_5\cdot NH\cdot NH_2 \longrightarrow \begin{array}{l}\text{CH=N}\cdot NH\cdot C_6H_5 \\ | \\ \text{C=N}\cdot NH\cdot C_6H_5 \\ | \\ \text{R}\end{array} + 2\,H_2O$$

Dissolve the carbohydrate (1 g) in water (5 cm³) and add phenylhydrazine (1 cm³) and ethanoic (acetic) acid (1 cm³). Heat the mixture on a boiling-water bath for 30 min and allow to cool. Filter off the product, wash well with cold water and recrystallize from ethanol.

CARBOXYLIC ACIDS

(a) Amide, anilide and 4-toluidide

Attempts to convert hydroxy- or amino-acids into amides via the acid chloride may result in the formation of undesired products. This may be avoided by using dicyclohexylcarbodi-imide (DCC) as in (ii) below.

49

$$\text{(i)} \quad RCO_2H + SOCl_2 \longrightarrow RCOCl \begin{cases} \xrightarrow{NH_3} & RCO \cdot NH_2 \\ \xrightarrow{C_6H_5 \cdot NH_2} & RCO \cdot NH \cdot C_6H_5 \\ \xrightarrow{4\text{-}CH_3 \cdot C_6H_4 \cdot NH_2} & RCO \cdot NH \cdot C_6H_4 \cdot 4\text{--}CH_3 \end{cases}$$

Moisture must be **excluded** until the base has been added.

To the acid (1 g), add thionyl chloride (2 cm³) and dimethylmethanamide (dimethylformamide, 2 drops) and heat under reflux on a boiling-water bath until no further reaction occurs (about 30 min). Distil off the excess of thionyl chloride and add the residue to a well stirred excess of ammonia (0.880, 10 cm³), or of aniline (1 cm³) or 4-toluidine (1 g) in dry ether (10 cm³).

Filter off the solid product and recrystallize from water, aqueous ethanol or ethanol.

$$\text{(ii)} \quad RCO_2H + R'NH_2 + C_6H_{11}N{=}C{=}NC_6H_{11} \longrightarrow RCONHR' +$$

$$C_6H_{11}NH \cdot CO \cdot NH \cdot C_6H_{11}$$

Add DCC (0.8 g) to a mixture of the carboxylic acid (0.8 g) and either aniline (1 cm³) or 4-toluidine (1 g) in trichloromethane (chloroform) or tetrahydrofuran (20 cm³). Shake or stir well and keep the mixture at 40–50° for 5 min. Allow to cool and stand for 30 min. Filter off the solid which is a mixture of substituted amide and dicyclohexylurea. Extract the urea by heating with ethanol (30 cm³) for about 5 min. The residue may be dried or recrystallized from water, aqueous ethanol or methanol.

(b) 4-Bromophenacyl ester and 4-phenylphenacyl ester

$$RCO_2H + ArCO \cdot CH_2Br \longrightarrow RCO \cdot O \cdot CH_2 \cdot COAr + HBr$$

where Ar = $4\text{-}BrC_6H_4-$ or $4\text{-}C_6H_5 \cdot C_6H_4-$.

Prepare a solution of the acid (1 g) in an equivalent amount of sodium hydroxide solution, make *slightly* acid to litmus by adding a few drops of 2 M hydrochloric acid and add the phenacyl bromide (1 g) (CAUTION: these bromides are eye and skin irritants) in ethanolic solution. Heat to boiling, adding more ethanol if solution is not complete. Continue refluxing for 1, 2 or 3 hours depending on whether the acid is mono-, di- or tri-basic. Cool and filter the product. Recrystallize from ethanol, aqueous ethanol or light petroleum (60–80°).

ENOLS

Semicarbazone and 2,4-dinitrophenylhydrazone

$$\underset{\underset{OH}{|}}{RC}{:}CHR' \rightleftharpoons \underset{\underset{O}{\|}}{RC} \cdot CH_2R' \xrightarrow{H_2N \cdot NH \cdot CO \cdot NH_2} \underset{\underset{N \cdot NH \cdot CO \cdot NH_2}{\|}}{RC} \cdot CH_2R' + H_2O$$

$$\Big\downarrow 2,4\text{-}(NO_2)_2 \cdot C_6H_3 \cdot NH \cdot NH_2$$

$$\underset{\underset{2,4\text{-}(NO_2)_2 C_6H_3 \cdot NH \cdot N}{\|}}{RC} \cdot CH_2R'$$

Prepare as described under *Aldehydes* (p. 44).

50

ESTERS

(a) Hydrolysis

$$RCO_2R' + H_2O \xrightarrow{\text{HO}^-} RCO_2H + R'OH$$

Various methods are available for the hydrolysis of esters to the parent acid and alcohol or phenol depending on the ease of hydrolysis of the ester and the boiling point of R'OH. Three methods are described below which make use of potassium hydroxide in different solvents, viz., water, ethanol and di(2-hydroxyethyl)ether (diethylene glycol).

Aqueous alkali – Reflux the ester (5 g) with 30% aqueous potassium hydroxide (40 cm³) until hydrolysis is complete, normally indicated by a change in appearance or odour of the reaction mixture. If a solid separates (usually a sparingly soluble salt of the acid component), filter, wash well with water and characterize as the 4-bromo- or 4-phenylphenacyl ester as described under *Carboxylic Acids* (p. 49). If a liquid separates, this may be an insoluble alcohol. In such cases, extract with ether, dry the ether extract over anhydrous sodium sulphate and evaporate the ether on a water bath. Characterize the compound obtained as described under *Alcohols* (p. 43).

Where a homogeneous solution is obtained, saturate with solid potassium carbonate and extract with ether. Dry the ether extract over anhydrous sodium sulphate and evaporate the ether. Characterize the residue as described under *Alcohols* (p. 43). Acidify the aqueous solution with hydrochloric acid and filter off any solid product. Wash well with water and characterize as described under either *Carboxylic Acids* (p. 49) or *Phenols* (p. 59), see Note 1. If no solid was obtained, saturate with calcium chloride and extract with ether. Evaporate the ether and characterize the residue as described under *Carboxylic Acids* (p. 49) or *Phenols* (p. 59), see Note 1.

Methanolic alkali – (for esters of high-boiling alcohols and phenols). Reflux the ester (5 g) with 20% methanolic potassium hydroxide (40 cm³) until hydrolysis is complete. If a solid separates, filter off, wash well with methanol and characterize as described under *Carboxylic Acids* (p. 49) or phenols (p. 59), see Note 1. Carefully distil the excess methanol from the filtrate and characterize the residue as described under *Alcohols* (p. 43) or *Phenols* (p. 59). If a homogeneous solution is obtained after hydrolysis, distil off the bulk of the methanol on a water bath, cool and extract the residue with ether. Dry the ether extract over anhydrous sodium sulphate, evaporate the ether and characterize the residue as described under *Alcohols* (p. 43). Characterize the ether-insoluble residue remaining as described under *Carboxylic Acids* (p. 49).

Alkali in diethylene glycol – (for esters resistant to hydrolysis). To the ester (5 g) add a solution consisting of potassium hydroxide (2 g) in diethylene glycol (10 cm³) and water (2 cm³). Reflux the mixture for 5 min. Distil off all the volatile material (water and alcohol) and saturate this distillate with potassium carbonate before extracting with ether. Dry the ether extract, distil off the solvent using a fractionating column and characterize the residue as described under *Alcohols* (p. 43). If no alcohol is detected, the residue remaining after ether extraction will contain a carboxylic acid and phenol. Dissolve in water and acidify with hydrochloric acid. Filter off any solid obtained and characterize as described under *Carboxylic Acids* (p. 49) or

Phenols (p. 59), see Note 1. The filtrate will contain the remaining component which should be characterized in the same manner.

ETHERS

(a) Nitro derivative

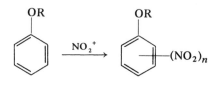

where $n = 1$, 2 or 3.

Nitration is sometimes a hazardous procedure and must always be undertaken with caution. Various nitration methods are available depending on the ease with which the particular compound may be nitrated. These methods are described below and the preferred method is indicated in the melting point tables.

(i) Mix equal volumes of concentrated sulphuric acid and concentrated nitric acid (5 cm^3) and add the organic compound (0.5 g). Maintain the temperature at 25° by cooling and shaking until the reaction is complete. If no reaction occurs, the mixture should be heated cautiously to start the reaction. Pour the resulting material into water (50 cm^3) and stir. Filter off the solid which is formed.

(ii) Mix concentrated sulphuric acid (5 cm^3) and fuming nitric acid (3 cm^3) (CAUTION) and cool to room temperature. Add the organic compound (0.5 g) with continuous shaking and cooling. When the initial reaction has subsided, heat for 5 min on a boiling-water bath. Pour into cold water (50 cm^3) and filter off the solid product. Sometimes an oil is obtained, but this will usually solidify on vigorous stirring and scratching.

(iii) Treat the organic compound (0.5 g) dropwise with fuming nitric acid (3 cm^3) with continuous cooling in ice. When the reaction subsides, allow the mixture to stand at room temperature for 5 min before pouring it into water (50 cm^3). Filter off the solid obtained.

(iv) Dissolve the organic compound (0.5 g) in the minimum of glacial acetic acid and add a mixture of fuming nitric acid (2 cm^3) and ethanoic (acetic) acid (2 cm^3). Heat the mixture to boiling and allow to stand until cold. Pour into water (50 cm^3) and filter off the resulting solid.

(v) As described in (iv) above but keep the mixture at 20° by cooling in ice. After standing for 5 min, dilute with water and filter off the solid product.

In all the above cases, the crude product should be thoroughly washed with water and recrystallized from aqueous ethanol, ethanol or benzene.

(b) Alkyl 3,5-dinitrobenzoate

$$ROR + 3,5\text{-}(NO_2)_2 C_6 H_3 \cdot COCl \longrightarrow 3,5\text{-}(NO_2)_2 C_6 H_3 \cdot CO \cdot OR + RCl$$

Treat the alcohol-free ether (1 g) with powdered anhydrous zinc chloride (0.1 g) and

3,5-dinitrobenzoyl chloride (0.5 g). Reflux gently for 1 hr, pour the product into saturated aqueous sodium carbonate solution (10 cm³) and heat on a boiling-water bath for 1 min. Allow to cool and filter off the solid obtained. Wash with sodium carbonate solution and then water. Dry the solid and extract it with boiling carbon tetrachloride. Evaporate the excess of solvent and allow the derivative to crystallize.

(c) Picric acid complex

$$ArOR + 2,4,6\text{-}(NO_2)_3C_6H_2{\cdot}OH \longrightarrow [ArOR]\,[2,4,6\text{-}(NO_2)_3C_6H_2{\cdot}OH]$$

Prepare as described under *Amines, Primary and Secondary* (p. 46).

(d) Sulphonamide

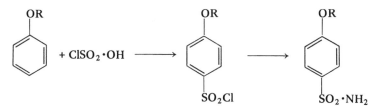

Prepare as described under *Halides, Aryl* (p. 55).

(e) Bromination

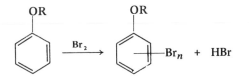

where $n = 1$, 2 or 3.

Suspend or dissolve the ether (1 g) in ethanoic (acetic) acid, trichloromethane (chloroform) or carbon tetrachloride (5 cm³) and add dropwise a solution of bromine in the same solvent until the colour of the bromine persists. Allow to stand for up to 15 min, adding more bromine solution if the colour fades. Evaporate the solvent (when using ethanoic (acetic) acid, pour into water) and recrystallize the product from ethanol.

HALIDES, ALKYL MONO-

(a) Thiouronium picrate

$$AlkX + H_2N{\cdot}CS{\cdot}NH_2 \longrightarrow \left[\begin{array}{c} NH_2 \\ \| \\ AlkS{\cdot}C{\cdot}NH_2 \end{array} \right]^{+} X^{-} \xrightarrow{\ 2,4,6\text{-}(NO_2)_3C_6H_2{\cdot}OH\ }$$

$$\left[\begin{array}{c} NH_2 \\ \| \\ AlkS{\cdot}C{\cdot}NH_2 \end{array} \right]^{+} [2,4,6\text{-}(NO_2)_3C_6H_2O]^{-} + HX$$

Treat the halide (1 cm³) with a solution of thiourea (1.5 g) in water (4 cm³) and ethanol

(3 cm^3). Heat on a boiling-water bath until solution is complete and then for a further 15 min. Pour the resulting solution into an excess of 1% aqueous picric acid solution and filter off the precipitate which forms. Recrystallize from aqueous ethanol.

(b) 2-Naphthyl ether

$$AlkX + 2\text{-}C_{10}H_7O^- \longrightarrow 2\text{-}C_{10}H_7 \cdot OAlk + X^-$$

Prepare a mixture of the halide (1 g), potassium hydroxide (1 g) and 2-naphthol (2 g) in ethanol (10 cm^3). Boil under reflux for 15 min and add a further portion of potassium hydroxide (2 g) in water (20 cm^3). Shake until a solid product is obtained, filter off and wash well with water. Recrystallize from aqueous ethanol. Isolate liquid products by ether extraction.

(c) 2-Naphthyl ether picrate

$$2\text{-}C_{10}H_7 \cdot OAlk + 2,4,6\text{-}(NO_2)_3C_6H_2 \cdot OH \longrightarrow [2\text{-}C_{10}H_7 \cdot OAlk][2,4,6\text{-}(NO_2)_3C_6H_2 \cdot OH]$$

To the product (0.5 g) from (b) above in ethanol (2 cm^3) add cold saturated alcoholic picric acid solution (5 cm^3). Collect the precipitate formed and recrystallize from ethanol.

(d) Oxidation of substituted benzyl halides to the corresponding carboxylic acids

$$ArCH_2X \xrightarrow{[O]} ArCO_2H$$

(i) Chromic acid oxidation

Carry out as described under *Alcohols* (p. 43).

(ii) Potassium permanganate oxidation

Mix the halide (1.5 g) with sodium hydroxide (1 g) and potassium permanganate (9 g) in water (100 cm^3). Reflux the solution until the colour of the permanganate is discharged and then filter off the manganese dioxide formed. Acidify the filtrate with concentrated hydrochloric acid and filter off the solid obtained. Recrystallize from water, aqueous ethanol or ethanol.

HALIDES, ALKYL POLY-

(a) Thiouronium picrate, e.g.

$$ClCH_2 \cdot CH_2Cl + 2H_2N \cdot CS \cdot NH_2 \xrightarrow{\text{picric acid}}$$

$$\left[\begin{array}{c} H_2N \cdot C \cdot S \cdot CH_2 \cdot CH_2 \cdot S \cdot C \cdot NH_2 \\ \| \qquad\qquad\qquad\qquad \| \\ NH_2 \qquad\qquad\qquad\quad NH_2 \end{array} \right]^{2+} 2[2,4,6\text{-}(NO_2)_3C_6H_2O]^-$$

Prepare as described under *Halides, Alkyl Mono-* (p. 53).

(b) 2-Naphthyl ether, e.g.

$$CH_2Cl_2 + 2[2\text{-}C_{10}H_7O^-] \longrightarrow [2\text{-}C_{10}H_7O]_2CH_2 + 2Cl^-$$

Prepare as described under *Halides, Alkyl Mono-* (see above).

HALIDES, ARYL

(a) Sulphonamide

$$ArX + ClSO_2 \cdot OH \xrightarrow[\text{excess}]{} 4\text{-}XAr \cdot SO_2 Cl \xrightarrow{NH_3} 4\text{-}XAr \cdot SO_2 \cdot NH_2$$

Prepare a solution of the compound (1 g) in dry trichloromethane (chloroform) (5 cm³), cool in ice and add chlorosulphonic acid (3 cm³). When the evolution of hydrogen chloride slackens, warm and maintain at room temperature for 30 min (50° for 10 min if reaction is slow). Pour the product into crushed ice, separate the trichloromethane layer, dry over anhydrous sodium sulphate and evaporate the trichloromethane on a boiling-water bath. Add to the residue ammonia solution (0.88, 10 cm³), boil for 10 min (fume cupboard), cool and dilute with water (10 cm³). Filter off the crude sulphonamide and recrystallize from aqueous ethanol.

(b) Nitro-derivative

$$ArX \xrightarrow{NO_2^+} 4\text{-}XAr \cdot NO_2$$

Prepare as described under *Ethers* (p. 52).

(c) Picric acid complex

$$ArX + 2,4,6\text{-}(NO_2)_3 C_6 H_2 \cdot OH \longrightarrow [ArX][2,4,6\text{-}(NO_2)_3 C_6 H_2 \cdot OH]$$

Prepare as described under *Amines, Primary and Secondary* (picrate) (p. 46).

HYDRAZINE DERIVATIVES

(a) Benzoyl derivative

$$RNH \cdot NH_2 + C_6 H_5 COCl \longrightarrow RNH \cdot NH \cdot CO \cdot C_6 H_5 + HCl$$

Prepare as described under *Amines, Primary and Secondary* (p. 46).

(b) Hydrazone from hydrazine derivative

$$RR'N \cdot NH_2 + C_6 H_5 \cdot CHO \longrightarrow RR'N \cdot N{:}CH \cdot C_6 H_5 + H_2 O$$

Dissolve the hydrazine derivative (0.5 g) in ethanoic (acetic) acid (1 cm³) and water (1 cm³). To this solution add benzaldehyde (0.5 g) and warm for 5 min. Add water (5 cm³) and filter the solid obtained. Recrystallize from ethanol or ethanoic acid.

HYDROCARBONS

(a) Diels–Alder adduct with maleic anhydride or benzoquinone, e.g.

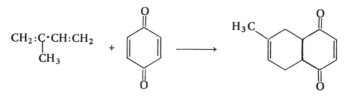

Dissolve the hydrocarbon (1 g) in xylene (5–10 cm³) and add powdered maleic anhydride (1 g) or benzoquinone (1 g). Reflux the mixture for 25 min and cool. If no solid separates, add small quantities of light petroleum (60–80°) to precipitate the product. Filter off the solid obtained and wash with a little light petroleum before recrystallizing from methanol, cyclohexane or xylene.

(b) Sulphonamide

$$ArH + ClSO_2OH \xrightarrow[\text{excess}]{} ArSO_2Cl \xrightarrow{NH_3} ArSO_2 \cdot NH_2$$

Prepare as described under *Halides, Aryl* (p. 55).

(c) Mercury derivative of alkynes

$$2RC\vdots CH + K_2HgI_4 \longrightarrow (RC\vdots C)_2Hg$$

Dissolve mercury(II) chloride (6.6 g) in a solution of potassium iodide (16.3 g) in water (16.3 cm³) and add 2 M sodium hydroxide (12.5 cm³). Dissolve the alkyne (0.5 g) in ethanol (10 cm³) and add it dropwise to the prepared solution (10 cm³); filter off the precipitate immediately and wash with 50% aqueous ethanol. Recrystallize the product from ethanol or benzene.

(d) Picric acid and styphnic acid derivatives

Prepare as described under *Amines, Primary and Secondary* (picrate) (p. 46).

KETONES

(a) 2,4-Dinitrophenylhydrazone

$$RR'C:O + 2,4-(NO_2)_2C_6H_3 \cdot NH \cdot NH_2 \longrightarrow 2,4-(NO_2)_2C_6H_3 \cdot NH \cdot N:CRR' + H_2O$$

Prepare as described under *Aldehydes* (p. 44).

(b) Semicarbazone

$$RR'C:O + H_2N \cdot NH \cdot CO \cdot NH_2 \longrightarrow RR'C:N \cdot NH \cdot CO \cdot NH_2 + H_2O$$

Prepare as described under *Aldehydes* (p. 44).

(c) Oxime

$$RR'C:O + H_2N \cdot OH \longrightarrow RR'C:N \cdot OH + H_2O$$

Dissolve the compound (0.5 g) in ethanol (3 cm³) and water (1 cm³) and add hydroxylamine hydrochloride (0.3 g) followed by sodium hydroxide (0.5 g). When solution is complete, reflux for 5–10 min. Cool in ice and acidify with hydrochloric acid (use litmus paper). Filter off the product and recrystallize from ethanol.

(d) 4-Nitrophenylhydrazone and phenylhydrazone

$$RR'C{:}O + ArNH{\cdot}NH_2 \longrightarrow RR'C{:}N{\cdot}NHAr + H_2O$$

where $Ar = 4\text{-}NO_2{\cdot}C_6H_4-$ or C_6H_5-.

Prepare as described under *Aldehydes* (p. 45).

(e) Benzylidene derivative

$$
\begin{array}{l}
RCH_2 \\
| \\
C{:}O + 2C_6H_5{\cdot}CHO \\
| \\
R'CH_2
\end{array}
\longrightarrow
\begin{array}{l}
RC{:}CHC_6H_6 \\
| \\
C{:}O \qquad + 2H_2O \\
| \\
R'C{:}CH{\cdot}C_6H_5
\end{array}
$$

Shake a mixture of the compound (0.5 g), benzaldehyde (1.2 cm^3) in a little ethanol and 4 M sodium hydroxide (0.5 cm^3). Allow to stand at room temperature until a crystalline product is obtained. Scratching the vessel with a glass rod will often induce crystallization. Filter off the product and recrystallize from ethanol.

NITRILES

(a) Amide

$$RCN \xrightarrow{\ H_2O_2\ } RCO{\cdot}NH_2$$

Prepare a solution containing 20 volume hydrogen peroxide (10 cm^3) and 2 M sodium hydroxide (2 cm^3), add the nitrile (0.5 g) and heat to 40° on a water bath. Shake frequently and finally filter off the solid product obtained. Wash and recrystallize from water, aqueous ethanol or ethanol.

(b) Nitro-derivative (for aromatic nitriles only)

$$ArCN \xrightarrow{\ NO_2^+\ } 3\text{-}NO_2{\cdot}Ar{\cdot}CN$$

Prepare as described under *Ethers*, Method (i) (p. 52).

(c) Hydrolysis to carboxylic acid

$$RCN \longrightarrow RCO_2H$$

Reflux the nitrile (1 g) with either 8 M sodium hydroxide (5 cm^3) for aliphatic nitriles or 7 M sulphuric acid (5 cm^3) for aromatic nitriles for 1 hr. Cool the resulting solution and add excess hydrochloric acid (aliphatic nitriles) or water (aromatic nitriles). Characterize the aliphatic acid produced as its 4-bromophenacyl ester as described under *Carboxylic Acids* (p. 49).

Aromatic acids produced may be filtered off and recrystallized from water, aqueous ethanol or ethanol.

57

NITRO-, HALOGENONITRO-COMPOUNDS AND NITRO-ETHERS

(a) Nitration

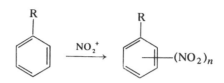

Prepare as described under *Ethers* (p. 52).

(b) Benzylidene derivative

$$RCH_2 \cdot NO_2 + C_6H_5 \cdot CHO \longrightarrow \underset{\underset{CH \cdot C_6H_5}{\|}}{RC \cdot NO_2} + H_2O$$

Prepare as described under *Ketones* (p. 56).

(c) Oxidation of alkyl side-chain to carboxylic acid

$$ArCH_3 \xrightarrow{[O]} ArCO_2H$$

Prepare as described under *Alcohols* (p. 43).

(d) Reduction to amine

$$RNO_2 \xrightarrow{[H]} RNH_2$$

Suspend the nitro-compound (1 g) in concentrated hydrochloric acid (10 cm³) and add ethanol (2 cm³) and tin (3 g). Cool until the initial reaction subsides and then heat under reflux for 30 min. Filter the solution, cool the filtrate and basify with 5 M sodium hydroxide, adding sufficient alkali to dissolve the precipitate tin(II) hydroxide formed. Extract the free amine with ether, dry the ether extract over anhydrous sodium sulphate and evaporate the ether (CARE). Further conversion to crystalline derivatives should be carried out as described under *Amines, Primary and Secondary* (p. 46).

(e) Partial reduction, e.g.

$$1,3\text{-}(NO_2)_2 C_6H_4 \longrightarrow 3\text{-}NO_2 \cdot C_6H_4 \cdot NH_2$$

Dissolve the nitro-compound (1 g) in ethanol (10 cm³) and add ammonia solution (0.88, 1 cm³). Saturate the cold solution with hydrogen sulphide and reflux on a boiling-water bath for 30 min. Cool, resaturate with hydrogen sulphide and reflux for a further 30 min. Pour into cold water and filter the solid obtained. Extract the solid with 2 M hydrochloric acid, basify this extract with ammonia solution (0.88) and filter the resulting nitro-amine. Recrystallize from aqueous ethanol, ethanol or benzene.

PHENOLS

(a) Ethanoate (acetate)

$$ArOH + CH_3 \cdot COCl \longrightarrow ArO \cdot CO \cdot CH_3 + HCl$$

(i) Dissolve the phenol (0.5 g) in dry pyridine (0.5 cm^3) and add ethanoyl (acetyl) chloride (0.5 cm^3) dropwise. Shake well after each addition and cool if the temperature rises rapidly. When the addition of ethanoyl chloride is complete, heat to 50–60° for 5 min. Cool, pour into water (15 cm^3) and stir until a solid is obtained. Filter off the solid and recrystallize from ethanol or aqueous ethanol.

(ii) Stir the phenol (0.5 g) with ethanoic (acetic) anhydride (0.5 cm^3) containing concentrated sulphuric acid (3 drops) at 60° for 15 min. Cool, add water (7 cm^3), stir well, and recrystallize the product.

(b) Benzoate and toluene-4-sulphonate

$$ArOH + C_6H_5 \cdot COCl \longrightarrow ArO \cdot CO \cdot C_6H_5 + HCl$$

$$ArOH + 4\text{-}CH_3 \cdot C_6H_4 \cdot SO_2 Cl \longrightarrow 4\text{-}CH_3 \cdot C_6H_4 \cdot SO_2 \cdot OAr + HCl$$

Prepare as described under *Alcohols* (p. 43). For nitrophenols it is preferable to replace this Schotten–Baumann method by one in which pyridine is the base, as described under *Alcohols*, method c (p. 43).

(c) 4-Nitrobenzoate and 3,5-dinitrobenzoate

$$ArOH + Ar'COCl \longrightarrow ArO \cdot COAr' + HCl$$

where Ar' = 4-NO$_2 \cdot$C$_6$H$_4$– or 3,5-(NO$_2$)$_2$C$_6$H$_3$–.
Prepare as described under *Alcohols*, method c (p. 43).

(d) Aryloxyethanoic acid

$$ArOH + ClCH_2 \cdot CO_2 H \longrightarrow ArO \cdot CH_2 \cdot CO_2 H + HCl$$

To a solution of the phenol (0.5 g) in 5 M sodium hydroxide (3 cm^3) add chloroethanoic (chloroacetic) acid (0.5 g) (CAUTION: this acid must not be allowed to come into contact with the skin). Add a little water if any solid is formed in the hot solution. Heat on a boiling-water bath for 1 hr, cool, acidify with 2 M hydrochloric acid to pH 3 and extract with ether. Extract the ethereal layer with 2 M sodium carbonate solution. If the sodium salt of the acid separates, remove by filtration and treat the solid with 2 M hydrochloric acid. The resulting solid is the required derivative. If no solid separates, acidify the sodium carbonate extract with 2 M hydrochloric acid. Filter off the solid obtained. Recrystallize the product from water, aqueous ethanol or ethanol.

(e) Derivative for picric and styphnic acids, e.g.

$$2,4,6\text{-}(NO_2)_3 C_6H_2 \cdot OH + C_{10}H_8 \longrightarrow [2,4,6\text{-}(NO_2)_3 C_6H_2 \cdot OH][C_{10}H_8]$$

Prepare saturated solutions of either picric or styphnic acid and naphthalene in ethanol and mix. Warm gently for a few minutes and cool. A crystalline product is readily obtained which may be recrystallized from ethanol.

QUINONES

(a) Oxime, e.g.

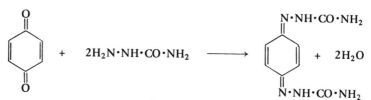

$$+ \quad 2H_2N \cdot OH \quad \longrightarrow \quad + \quad 2H_2O$$

Prepare as described under *Ketones* (p. 56).

(b) Semicarbazone, e.g.

$$+ \quad 2H_2N \cdot NH \cdot CO \cdot NH_2 \quad \longrightarrow \quad + \quad 2H_2O$$

Prepare as described under *Aldehydes* (p. 44).

(c) Quinol, e.g.

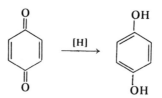

$$\xrightarrow{[H]}$$

Dissolve or suspend the quinone (1 g) in benzene (5–10 cm³) and treat with a solution (20 cm³) of sodium hydrosulphite (10%) in M sodium hydroxide. Shake until the quinone colour has disappeared and separate the aqueous layer. Cool this (ice) and acidify with concentrated hydrochloric acid. Filter off the solid obtained and recrystallize from water or ethanol.

SULPHONIC ACIDS AND THEIR DERIVATIVES

(a) Amide

$$RSO_2 \cdot OH + PCl_5 \quad \longrightarrow \quad RSO_2Cl \quad \xrightarrow{NH_3} \quad RSO_2 \cdot NH_2$$

Mix the dry sulphonic acid (1 g) or the dry sodium salt (1 g) with phosphorus pentachloride (2 g) and heat on a boiling-water bath, taking care to exclude water vapour from the reaction vessel. When reaction has ceased, add water (15 cm³) and stir. Decant the water and add ammonia solution (0.88, 3 cm³) to the residue. Heat on a boiling-water bath for 5–10 min and cool. Filter off the solid product, wash well with water and recrystallize from water or aqueous ethanol.

(b) Anilide

$$RSO_2 \cdot OH + PCl_5 \longrightarrow RSO_2 Cl \xrightarrow{C_6 H_5 \cdot NH_2} RSO_2 \cdot NH \cdot C_6 H_5$$

Use the method described above but replace the ammonia solution with aniline (1 cm³).

(c) S-Benzylisothiouronium salt

$$RSO_2 \cdot O^- Na^+ + \begin{bmatrix} C_6 H_5 \cdot CH_2 \cdot S \cdot C:NH_2 \\ | \\ NH_2 \end{bmatrix}^+ Cl^- \longrightarrow$$

$$\begin{bmatrix} C_6 H_5 \cdot CH_2 \cdot S \cdot C:NH_2 \\ | \\ NH_2 \end{bmatrix}^+ [O \cdot SO_2 R]^- + NaCl$$

Prepare the sodium salt of the sulphonic acid (0.5 g) in water (3 cm³) by addition of 2 M sodium hydroxide until the solution is just alkaline to phenolphthalein. Neutralize the excess of alkali with a further addition of the sulphonic acid (or 2 M hydrochloric acid) and add a solution of S-benzylisothiouronium chloride (2 g) in water (5 cm³). Cool in ice, filter off the crystalline product and recrystallize from water, aqueous ethanol or ethanol.

(d) Xanthyl derivative of sulphonamides

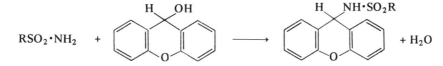

$$RSO_2 \cdot NH_2 \quad + \qquad \qquad \longrightarrow \qquad + H_2O$$

Treat the sulphonamide (0.5 g) with xanthydrol (0.5 g) in ethanoic (acetic) acid (25 cm³) and heat the mixture until solution is complete. Allow to stand at room temperature until a solid separates. Water may be added if no solid separates. Filter off the product and recrystallize from aqueous dioxan.

(e) Benzoyl derivative of sulphonamides

$$RSO_2 \cdot NH_2 + C_6 H_5 COCl \longrightarrow RSO_2 \cdot NH \cdot CO \cdot C_6 H_5 + HCl$$

Prepare as described under *Alcohols* (p. 43).

(f) Ethanoyl (acetyl) derivative of sulphonamides

$$RSO_2 \cdot NH_2 + CH_3 \cdot COCl \longrightarrow RSO_2 \cdot NH \cdot CO \cdot CH_3 + HCl$$

To the sulphonamide (1 g) add ethanoyl (acetyl) chloride (3 cm³) and reflux for 30 min, adding ethanoic (acetic) acid (up to 2 cm³) if solution is not complete. Remove the excess of ethanoyl chloride by vacuum distillation and pour the residue into ice cold water (25 cm³). Stir the product until it solidifies, filter off, wash well with water and recrystallize from aqueous ethanol.

(g) Diphenylmethyl derivative

$$RSO_2NH_2 + Ph_2CHOH \longrightarrow RSO_2NHCHPh_2 + H_2O$$

Prepare as described under *Amides, Imides, Ureas and Guanidines* (p. 45).

THIOETHERS (SULPHIDES)

(a) Sulphone

$$RSR' \xrightarrow{[O]} RSO_2R'$$

Dissolve the thioether (1 g) in the minimum of ethanoic (acetic) acid and add 3% potassium permanganate solution as long as the colour is discharged. If starting material precipitated during this addition, more ethanoic acid should be added. When reaction is complete, pass in sulphur dioxide until the manganese dioxide precipitate has just dissolved. Add crushed ice and filter off the solid sulphone. Wash well with water and recrystallize from ethanol.

THIOLS AND THIOPHENOLS

(a) 2,4-Dinitrophenylsulphide

$$RSH + 2,4\text{-}(NO_2)_2C_6H_3Cl \longrightarrow 2,4\text{-}(NO_2)_2C_6H_3 \cdot SR + HCl$$

Dissolve the thiol (1 g) in ethanol (30 cm³) and add sodium hydroxide (0.4 g) in ethanol (3 cm³) followed by 1-chloro-2,4-dinitrobenzene (2 g) (CAUTION—skin irritant) in ethanol (10 cm³). Reflux on a boiling-water bath for 10 min, filter and allow the filtrate to cool. Recrystallize the product from ethanol.

(b) Hydrogen 3-nitrophthaloyl derivative

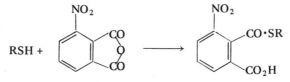

Prepare as described under *Alcohols* (p. 43).

(c) 3,5-Dinitrobenzoyl derivative

$$RSH + 3,5\text{-}(NO_2)_2C_6H_3 \cdot COCl \longrightarrow 3,5\text{-}(NO_2)_2C_6H_3 \cdot CO \cdot SR + HCl$$

Prepare as described under *Alcohols* (p. 43).

6

TABLES OF ORGANIC COMPOUNDS
AND THEIR DERIVATIVES

EXPLANATORY NOTES ON THE TABLES
OF COMPOUNDS AND DERIVATIVES

1. In each table the compounds are listed in the order of increasing boiling point if they are liquids or solids melting below 40°. Compounds which melt at 40° or above are divided from the liquids by a horizontal line and are arranged in the order of increasing melting point although the boiling point is sometimes also included.

2. Boiling points are given at atmospheric pressure except for a few high boiling compounds whose boiling points are given at reduced pressure and are written thus: 94/12 mm, which means a boiling point of 94° at 12 mm pressure.

3. Boiling and melting points are given to the nearest whole number and as one value only. This approximation is made for simplicity and because of slight variation in the degree of accuracy of different thermometers and in the personal element in determining the melting or boiling point. For example, a melting point which is recorded in the literature as 172–174° or as 173.5° is given in the tables as 173°.

4. The figures given in the columns of derivatives are melting points. Two different values of a boiling or melting point are recorded in the chemical literature for a few compounds. The second (usually less frequently encountered) value is given in the tables in parentheses below the more common value.

5. For compounds which exist as enantiomorphs, the constants are those of the racemic or (±)-modification unless otherwise stated.

6. The following abbreviations are used in the tables:

anhyd. anhydrous	*dil*. dilute
aq. aqueous	*hyd*. hydrate
conc. concentrated	*insol*. insoluble
d. decomposition	*sol*. soluble
deriv. derivative	*subl*. sublimes

7. When the colour of a compound is other than white it is given in the last column of the table which also contains supplementary information which may assist in the identification of the compound.

63

Table 1. Acetals

Acetal	B.p.	Aldehyde	4-Nitrophenyl-hydrazone (p. 45)	2,4-Dinitro-phenylhydra-zone (p. 44)	Alcohol	K alkyl xanthate (p. 43)
Dimethoxymethane (Methylal)	45	Methanal	182	167	Methanol	182
1,1-Dimethoxyethane (Dimethylacetal)	64	Ethanal	128	168	Methanol	182
Diethoxymethane (Ethylal)	89	Methanal	182	167	Ethanol	225
1,1-Diethoxyethane (Acetal)	102	Ethanal	128	168	Ethanol	225
1,1-Diethoxyprop-2-ene (Acrolein acetal)	126	Propenal (Acrolein)	151	165	Ethanol	225
1,1-Dipropoxymethane	140	Methanal	182	167	Propan-1-ol	233
1,1-Dibutoxyethane	187	Ethanal	128	168	Butan-1-ol	255
αα-Dimethoxytoluene	198	Benzaldehyde	192	237	Methanol	182
αα-Diethoxytoluene	222	Benzaldehyde	192	237	Ethanol	225

Table 2. Alcohols (C, H and O)

	B.p.	M.p.	3,5-Dinitro-benzoate (p. 43)	H 3-nitro-phthalate (p. 43)	4-Nitro-benzoate (p. 43)	Notes
Methanol	65		109	153*	96	*Anhydrous; monohydrate melts <100 but if dried at 80, it becomes anhydrous
Ethanol	78		94	157	56	
Propan-2-ol	83		122	154	110	
2-Methylpropan-2-ol (t-Butanol)	83	25	142		116	
Propan-1-ol	97		75	144	35	
Prop-2-enol (Allyl alcohol)	97		50	124	28	Unsaturated
Butan-2-ol (s-Butanol)	100		76	131	25	
2-Methylbutan-2-ol (t-Pentyl alcohol)	102		118		85	
2-Methylpropan-1-ol (Isobutanol)	108		88	183	69	
3-Methylbutan-2-ol	113		76	127		
Pentan-3-ol	116		100	121	17	
Butan-1-ol	117		64	147	35	
Pentan-2-ol	120		62	103	17	
3-Methylpentan-3-ol	123		96		69	
2-Methoxyethanol (Methyl cellosolve)	125			129	50	
2-Methylbutan-1-ol (Amyl alcohol)	128		70	158		
3-Methylbutan-1-ol	132		62	164	21	
4-Methylpentan-2-ol	132		65	166	26	
2-Ethoxyethanol (Ethyl cellosolve)	135		75	121*		*Anhydrous; monohydrate, 94
Hexan-3-ol	136		77	127		
Pentan-1-ol	138		46	136	oil	
2,4-Dimethylpentan-3-ol	140		38	151	40	
Cyclopentanol	140		115		62	

Table 2. Alcohols (C, H and O) (*cont.*)

	B.p.	M.p.	3,5-Dinitro-benzoate (p. 43)	H 3-nitro-phthalate (p. 43)	4-Nitro-benzoate (p. 43)	Notes
3-Hydroxybutan-2-one (Acetoin)	145					See Table 26
1-Hydroxypropan-2-one (Acetol)	146					See Table 26
2-Methylpentan-1-ol	148		50	145		
2-Ethylbutan-1-ol	149		51	147		
4-Methylpentan-1-ol	152		70	140		
Hexan-1-ol	156		61	124		
Heptan-2-ol	158		50			$K_2Cr_2O_7$—H_2SO_4 (p. 44) → heptan-2-one
Cyclohexanol	161	25	113	160	52	
Furfuryl alcohol	170		81		76	
2-Butoxyethanol (Butyl cellosolve)	171			120	oil	
Heptan-1-ol	176		48	127	oil	
Tetrahydrofurfuryl alcohol	177		83		47	
Octan-2-ol	179		32		28	$K_2Cr_2O_7$—H_2SO_4 (p. 44) → octan-2-one
2-Ethylhexan-1-ol	184			108		
Propane-1,2-diol (Propylene glycol)	187		147		127	
3,5,5-Trimethylhexan-1-ol	193		62	150		
Octan-1-ol	194		62	128	oil	
(−)-Linalyl alcohol	197		135		70	Unsaturated
Ethane-1,2-diol (Ethylene glycol)	197		169		140	Ditoluene-4-sulphonate, 93 (p. 43)
1-Phenylethanol	202	20	94		43	
Benzyl alcohol	205		113	183	85	
Nonan-1-ol	214		52	125		
Propane-1,3-diol (Trimethylene glycol)	214		178		119	Ditoluene-4-sulphonate, 93 (p. 43)
Isoborneol	216		138		129	
2-Phenylethanol	219		108	123	62	
α-Terpineol	221	35	78		139	
Tetradecan-1-ol (Myristic alcohol)	221	39	67	123	51	
trans-3,7-Dimethylocta-2,6-dien-1-ol (Geraniol)	229		63	117	35	Unsaturated; Br_2 → tetra-bromide, 70
Butane-1,4-diol	230	19			175	Dibenzoate, 81; ditoluene-4-sulphonate, 94 (p. 43)
Decan-1-ol	231	6	57	123	30	
2-Phenoxyethanol	237 (245)	12	105 (74)	113	63	Toluene-4-sulphonate, 80 (p. 43)
3-Phenylpropan-1-ol	237		92	117	46	
Undecan-1-ol	243	15	55	123	30	
Di(2-hydroxyethyl) ether (Diethylene glycol)	244		150		100 di	Ditoluene-4-sulphonate, 88 (p. 43)
3-Phenylprop-2-en-1-ol (Cinnamyl alcohol)	257	33	121		78	Unsaturated; Br_2 → dibromide, 74
Dodecan-1-ol (Lauryl alcohol)	259	25	60	124	43	
Propane-1,2,3-triol (Glycerol)	290d	18			188	

Table 2. Alcohols (C, H and O) (*cont.*)

	B.p.	M.p.	3,5-Dinitro-benzoate (p. 43)	H 3-nitro-phthalate (p. 43)	4-Nitro-benzoate (p. 43)	Notes
(−)-2-Isopropyl-5-methyl cyclohexanol ((−)-Menthol)	216	42	153		61	
Hexadecan-1-ol (Cetyl alcohol)		50	66	122	52 (58)	
Heptadecan-1-ol		54	121	121	53	
But-2-yn-1,4-diol		55	190			Dibenzoate, 76; di-4-toluenesulphonate, 94 (p. 43)
Octadecan-1-ol (Stearyl alcohol)		59	66	119	64	
Diphenylmethanol (Benzhydrol)		69	141		131	
D-Glucitol (D-Sorbitol)		111*				*Anhydrous; hydrate, 90. Hexaethanoate, 99; hexabenzoate, 129 (p. 43)
Benzoin		133			123	Benzoate, 124 (p. 43). See Table 26
Furoin		136				See Table 26
Triphenylmethanol		162				Ethanoate, 87; benzoate, 162 (p. 43)
D-Mannitol		166				Hexaethanoate, 126; hexabenzoate, 148(129) (p. 43)
D-Galactitol (Dulcitol)		188				Hexaethanoate, 171; hexabenzoate, 188 (p. 43)
(+)-Borneol	212	208	154		153	
Myo-inositol		225	86			Hexaethanoate, 212 subl.; hexabenzoate, 258 (p. 43)
Pentaerythritol		262 (253)				Tetraethanoate, 84; tetrabenzoate, 99 (p. 43)

Table 3. Alcohols (C, H, O and halogen or N)

	B.p.	M.p.	3,5-Dinitro-benzoate (p. 43)	H 3-nitro-phthalate (p. 43)	4-Nitro-benzoate (p. 43)	Notes
1-Chloropropan-2-ol	127		77			
2-Chloroethanol (Ethylene chlorohydrin)	129		95 (88)	98	56	
2-Chloropropan-1-ol	133		76			
2-(Dimethylamino)ethanol	135					See Table 12
2-Bromoethanol	149d		85	172		
2,2,2-Trichloroethanol	151	19	142		71	
2-(Diethylamino)ethanol	161					See Table 12
3-Chloropropan-1-ol	161		77			
1-Aminopropan-2-ol (Isopropanolamine)	163					Picrate, 142; see Table 8
2-Aminoethanol	171					Picrate, 159; reacts with phthalic anhydride → β-hydroxyethylimide, 127; see Table 8
3-Aminopropanol	188					See Table 8
Di-(2-hydroxyethyl)-amine (Diethanolamine)	270	28				See Table 11

Table 4. Aldehydes (C, H and O)

	B.p.	M.p.	2,4-Di-nitro-phenyl-hydra-zone (p. 44)	Semi-carba-zone (p. 44)	Di-meth-one (p. 45)	4-Nitro-phenyl-hydra-zone (p. 45)	Notes
Methanal (Formaldehyde)	−21		167	169d	189	182	40% aq. solution is formalin
Ethanal (Acetaldehyde)	20		168	163	140	128	
Propanal (Propionaldehyde)	50		155	154*	155	124	*Recryst. from water
Ethanedial (Glyoxal)	50		327	270	186* mono	310d	*Di, 228
Propenal (Acrolein)	52		165	171	192	151	Unsaturated
2-Methylpropanal (Isobutyraldehyde)	64		182	125	154	131	
2-Methylprop-2-enal	73		206	198			Unsaturated
Butanal	74		125	105	136	91	
2,2-Dimethylpropanal (Pivalaldehyde)	75		209	190		119	
3-Methylbutanal (Isovaleraldehyde)	92		123	132 (107)	155	110	
2-Methylbutanal	93		121	103			
Pentanal (Valeraldehyde)	103		107		105	74	
But-2-enal (Crotonaldehyde)	103		196	200*	186	184	*Varies with rate of heating. Unsaturated
5-Hydroxymethylfur-furaldehyde	114	35	184	195d (166)		185	
2-Ethylbutanal	116		134	98	102		
Paraldehyde	124						Gives ethanal on warming with trace of conc. sulphuric acid
Hexanal (Caproaldehyde)	129		104	108	109	80	
3-Methylbut-2-enal (β-Methylcrotonalde-hyde)	135		182	221			Unsaturated
Tetrahydrofurfural	144		134	166	123		
Heptanal (Heptaldehyde)	156		108	109	103 (135)	73	
Furfural	161		202*	203	160d	154	*Variable
Hexahydrobenzaldehyde	162		172	174			
Succinaldehyde	169		280				Oxime, 172 (p. 45)
Octanal	171		106	98	90	80	
Benzaldehyde	179		237	222*	195	192	*Varies with rate of heating. Smell of bitter almonds
Nonanal	185		100	100	86		
5-Methylfurfural	187		212	211		130	
Phenylethanal	194	33	121	156	165	151	
2-Hydroxybenzaldehyde (Salicylaldehyde)	196		248*	231	211	227	*From ethanol. See Table 30
3-Tolualdehyde	199		194	224	172	157	
2-Tolualdehyde	200		194	212	167	222	

Table 4. Aldehydes (C, H and O) (*cont.*)

	B.p.	M.p.	2,4-Di-nitro-phenyl-hydra-zone (p. 44)	Semi-carba-zone (p. 44)	Di-meth-one (p. 45)	4-Nitro-phenyl-hydra-zone (p. 45)	Notes
4-Tolualdehyde	204		234	234		200	
Decanal	208		104	102	92		
Phenoxyacetaldehyde	215d		130	145			Oxime, 95 (p. 45)
3-Phenylpropanal	224		149	127		123	
trans-3,7-Dimethylocta-2,6-dien-1-al (Geranial, citral a)	228d		116 (108)	164		195	Unsaturated
3-Methoxybenzaldehyde	230		218	233d		171	
4-Isopropylbenzaldehyde (Cuminaldehyde)	235		244	211	171	190	
2-Methoxybenzaldehyde	245	38	253	215	188	205	
4-Methoxybenzaldehyde (Anisaldehyde)	248		253d	210	145	160	
3-Phenylpropenal (Cinnamaldehyde)	252		255d	215	219 (213)	195	Unsaturated
3,4-Methylenedioxy-benzaldehyde (Piperonal)	263	37	266d	234	178	200	
1-Naphthaldehyde	292	34	254	221		234	
Tetradecanal (Myristaldehyde)	155/10 mm	23	108	107		95	
Hexadecanal (Palmitaldehyde)	200/29 mm	34	108	108		97	
Octadecanal (Stearaldehyde)	212/22 mm	38	110	108		101	
Dodecanal (Lauraldehyde)		44	106	106		90	
Phthalaldehyde		56					Phenylhydrazone, 191 (p. 45)
3,4-Dimethoxybenzal-dehyde (Veratraldehyde)		58	264	177	173		
2-Naphthaldehyde		60	270	245		230	
4-Hydroxy-3-methoxy-benzaldehyde (Vanillin)		80	271d	240d	197	227	Bisulphite addition compound is sol. in water. See Table 30
Phenylglyoxal, hydrate		91	296	217d		310	Dioxime, 168; mono-oxime, 128 (p. 45)
3-Hydroxybenzaldehyde		104	260	199		222	See Table 30
Terephthalaldehyde		116		225d		281*	*Sinters at 272. Oxime, 200 (p. 45)
4-Hydroxybenzaldehyde		117	280d	222	189	266	See Table 30; bisulphite compound is sol. in water
2,4-Dihydroxybenzaldehyde (β-Resorcylaldehyde)		135	286d (302)	260d	226		See Table 30
Glyceraldehyde (dimer)		142	167	160d	197		
3,4-Dihydroxybenzaldehyde (Protocatechualdehyde)		154	275d	230d	143d		See Table 30

Table 5. Aldehydes (C, H, O and halogen or N)

	B.p.	M.p.	2,4-Di-nitro-phenyl-hydra-zone (p. 44)	Semi-carba-zone (p. 44)	Di-meth-one (p. 45)	4-Nitro-phenyl-hydra-zone (p. 45)	Notes
Trichloroethanal (Chloral)	98		131	90d			
Tribromoethanal (Bromal)	174*						*Anhydrous; hydrate, 54. Oxime, 115 (p. 45)
2-Chlorobenzaldehyde	208	11	208	225	205	249	
3-Chlorobenzaldehyde	213	18	255	228		216	
2-Bromobenzaldehyde	230	22	203	214		240	
3-Bromobenzaldehyde	234		256	205		220	
2-Aminobenzaldehyde		40	250	247		219	Oxime, 135 (p. 45)
4-(Diethylamino)benzal-dehyde		41	206	214			Oxime, 93 (p. 45)
2-Nitrobenzaldehyde		44	250d	256		263	
4-Chlorobenzaldehyde	214	47	268	231		239	
Trichloroethanal hydrate (Chloral hydrate)		57	131	90d			
4-Bromobenzaldehyde		57	257* 128*	228		208	*Polymorphs
3-Nitrobenzaldehyde		58	293d	246	198	247	
2,4-Dichlorobenzal-dehyde		71	226			256	Oxime, 136 (p. 45)
4-(Dimethylamino)benzal-dehyde		74	237	222		182	
4-Nitrobenzaldehyde		106	320	220	190	249	
5-Bromo-2-hydroxy-benzaldehyde (5-Bromosalicylalde-hyde)		106	292	297d			Oxime, 126 (p. 45). See Table 31

Table 6. Amides (primary), imides, ureas, thioureas and guanidines

	M.p.	Xanthyl deriv. (p. 45)	Carboxylic acid (p. 46)	Notes
Methanamide (Formamide)	3	184		B.p. 193; diphenylmethyl deriv., 133 (p. 45)
Ethyl carbamate (Urethane)	49	169		
Methyl carbamate	54	193		

69

Table 6. Amides (primary), imides, ureas, thioureas and guanidines (*cont.*)

	M.p.	Xanthyl deriv. (p. 45)	Carboxylic acid (p. 46)	Notes
Propanamide	81	214		Diphenylmethyl deriv., 143 (p. 45)
Ethanamide (Acetamide)	82	240		Diphenylmethyl deriv., 148 (p. 45)
Propenamide (Acrylamide)	84			Unsaturated
2-Phenylpropanamide	92	158		
Maleimide	93		130	Unsaturated
Semicarbazide	96			Ethanal semicarbazone, 163
N-Methylurea	101	230		Ethanoyl deriv., 180
3-Phenylpropanamide	105	189	48	
Butanamide	116	186		Diphenylmethyl deriv., 122 (p. 45)
Chloroethanamide	120	209	*	*Hydrolysis gives hydroxyethanoic acid, 80
Cyanoethanamide	123	223		
Succinimide	125	246	185	
Benzamide	128	223	122	Diphenylmethyl deriv., 175 (p. 45)
Urea	132	274		Nitrate, 163
2-Hydroxybenzamide (Salicylamide)	139		158	See Table 31; diphenylmethyl deriv., 160 (p. 45)
N-Phenylurea	147	225		
N-Phenylthiourea	154			
Phenylethanamide	157	196	76	Diphenylmethyl deriv., 164 (p. 45)
Malonamide	170	270	133d	
Guanidine hydrochloride	172			Cold conc. HNO_3 + H_2SO_4 → nitroguanidine, 230d
4-Ethoxyphenylurea (Dulcin)	173			Heating above m.p. → di(4-ethoxyphenyl)urea, 235
Thiourea	180			Heating at m.p. → NH_4CNS which with aq. $FeCl_3$ gives red colour; N,S-diethanoyl deriv. 153; N-benzoyl deriv. 171
NN-Dimethylurea	182	250		
Thiosemicarbazide	182			Ethanoyl deriv., 165; benzaldehyde thiosemicarbazone, 160
1,1-Diphenylurea (Carbanilide)	189	180		
Biuret*	192d	260		*$H_2N \cdot CO \cdot NH \cdot CO \cdot NH_2$. With trace of $CuSO_4$ + dil. NaOH → red colour; an excess of $CuSO_4$ gives violet colour
Guanidine carbonate	197			Cold conc. HNO_3 + H_2SO_4 → nitroguanidine, 230d
4-Nitrobenzamide	201	232	240	Diphenylmethyl deriv., 223 (p. 45)
N-Cyanoguanidine (Dicyandiamide)	208			
Guanidine nitrate	214			Cold conc. HNO_3 + H_2SO_4 → nitroguanidine, 230d
Phthalamide	220		200d	Loses NH_3 near its m.p. to give phthalimide, 233
2-Benzoic sulphimide (Saccharin)	230	199		
Phthalimide	233	177	200d	
Succinamide	260d	275	125	Diphenylmethyl deriv., 221 (p. 45)

Table 7. Amides, *N*-substituted. *N*-Substituted amides listed as ethanoyl and benzoyl derivatives of amines and amino-acids in Tables 8–11 and 13 should be consulted together with the following list. Hydrolysis (p. 46) should be followed by identification of the acid and amine.

	B.p.		M.p.
NN-Dimethylmethanamide		*N*-(2,4-Dimethylphenyl)ethanamide	130
(*NN*-Dimethylformamide)	153	*N*-Phenylbromoethanamide	131
NN-Dimethylethanamide		*N*-(4-Ethoxyphenyl)ethanamide	
(*NN*-Dimethylacetamide)	165	(Phenacetin)	135[c]
NN-Diethylmethanamide		*N*-Phenyl-4-toluamide	144
(*NN*-Diethylformamide)	176	*N*-4-Tolylethanamide	153[d]
		N-Methyl-*N*-(4-nitrophenyl)ethanamide	153
	M.p.	*N*-4-Tolybenzamide	158
		N-1-Naphthylethanamide	160
N-Phenylmethanamide (Formanilide)	46	*N*-Phenylbenzamide (Benzanilide)	163[e]
3-Oxobutanamide (Acetoacetamide)	54	*N*-(4-Bromophenyl)ethanamide	
N-Ethyl-*N*-phenylethanamide		(4-Bromoacetanilide)	167
(*N*-Ethylacetanilide)	54[a]	*N*-Bromobutanedioic imide	
N-Phenyl-3-oxobutanamide		(*N*-Bromosuccinimide)	174
(Acetoacetanilide)	85	*N*-(4-Chlorophenyl)ethanamide	178
N-(2-Methoxyphenyl)ethanamide	87	*NN'*-Diethanoyl-1,2-diaminobenzene	186
N-Phenyloctadecanamide	94	*N*-Benzoylglycine (Hippuric acid)	187
NN-Diphenylethanamide	101	*NN'*-Diethanoyl-1,3-diaminobenzene	190
N-Methyl-*N*-phenylethanamide	102	*N*-(4-Hydroxyphenyl)-*N*-methylethanamide	240
N-Phenylpropanamide	106[b]	Barbituric acid	245
N-Phenylethanamide (Acetanilide)	114	*NN'*-Diethanoyl-1,4-diaminobenzene	303

[a]Conc.HNO_3 + H_2SO_4 at 40° → 4-nitro deriv., 118.
[b]Conc.HNO_3 + H_2SO_4 at 0° → 4-nitro deriv., 182.
[c]Warm 10% HNO_3 → 3-nitro deriv., 103.
[d]Br_2-ethanoic acid → 3-bromo deriv., 117.
[e]Br_2-ethanoic acid → 4-bromo deriv., 202.

Table 8. Amines, primary aliphatic

	B.p.	M.p.	Ethanoyl deriv. (p. 46)	Benzoyl deriv. (p. 46)	Toluene-4-sulphonyl deriv. (p. 46)	2,4-Dinitrophenyl deriv. (p. 47)	Notes
Methylamine	−7		28	80	77	178	Normally supplied as
Ethylamine	17			69	62	113	an aqueous solution
2-Aminopropane	32			100	51	95	
1,1-Dimethylethylamine	46			134	114		Picrate, 198 (p. 47)
(t-Butylamine)							
Propylamine	49			85	52	97	
Prop-2-enylamine	58			Oil	64	76	Picrate, 140 (p. 47)
(Allylamine)							
1-Methylpropylamine	63			76	55		Picrate, 140 (p. 47)
(s-Butylamine)							
2-Methylpropylamine	69			57	78	94	
(Isobutylamine)						(80)	
Butylamine	77			42	65	90	Picrate, 151 (p. 47)
3-Methylbutylamine	96				65	91	Picrate, 138 (p. 47)
(Isopentylamine)							

Table 8. Amines, primary aliphatic (*cont.*)

	B.p.	M.p.	Ethan-oyl deriv. (p. 46)	Benz-oyl deriv. (p. 46)	Tolu-ene-4-sulph-onyl deriv. (p. 46)	2,4-Di-nitro-phenyl deriv. (p. 47)	Notes
Pentylamine	104					81	Picrate, 139 (p. 47)
Ethylenediamine	116	8	172	249	160	306	
Propane-1,2-diamine (Propylenediamine)	119		139	192	103		
Hexylamine	130			40		39	Picrate, 127; benzene-sulphonamide, 96 (p. 46)
Cyclohexylamine	134		104	147	87	156	
Propane-1,3-diamine (Trimethylenediamine)	136		126	147	148		
Butane-1,4-diamine (Tetramethylenediamine)	159	27	137	177	224		
1-Aminopropan-2-ol (Isopropanolamine)	163					97	Picrate, 142; see Table 3
2-Aminoethanol	171			88 di	Oil	90	Picrate, 159; see Table 3
Benzylamine	184		60	106	116	116	
1-Phenylethylamine	185		57	120		118	
3-Aminopropanol	188				56	131	Picrate, 222; see Table 3
2-Phenylethylamine	197		51	116	64	154	
(−)-Menthylamine	205		145	156			Picrate, 215 (p. 47)
Dodecylamine	247	27			73	38	
Diphenylmethylamine (Benzhydrylamine)	303		146	172			
Hexane-1,6-diamine (Hexamethylenediamine)	204	42		155			Picrate, 220 (p. 47)

Table 9. Amines, primary aromatic (C, H, (O) and N)

	B.p.	M.p.	Ethan-oyl deriv. (p. 46)	Benz-oyl deriv. (p. 46)	Tolu-ene-4-sulph-onyl deriv. (p. 46)	2,4-Di-nitro-phenyl deriv. (p. 47)	Notes
Aniline	184		114	163	103	156	
2-Toluidine	200		109	144	110	120	
3-Toluidine	203		66	125	114	159	
2-Ethylaniline	210		112	147			Picrate, 194 (p. 47)
2,4-Dimethylaniline	212		130	192	181	156	
2,5-Dimethylaniline	213	15	139	140	119 (232)	150	
2,6-Dimethylaniline	215		177	168	212		
4-Ethylaniline	216		94	151	104		
2-Methoxyaniline (*o*-Anisidine)	218	5	85	60	127	151	
3,5-Dimethylaniline	220		144	136			

Table 9. Amines, primary aromatic (C, H, (O) and N) (*cont.*)

	B.p.	M.p.	Ethanoyl deriv. (p. 46)	Benzoyl deriv. (p. 46)	Toluene-4-sulphonyl deriv. (p. 46)	2,4-Dinitrophenyl deriv. (p. 47)	Notes
2,3-Dimethylaniline	221		134	189			
2-Ethoxyaniline (o-Phenetidine)	228		79	104	164	164	
2,4,6-Trimethylaniline	229		216	204	167		
3-Ethoxyaniline (m-Phenetidine)	248		96	103	157		
2-Aminoacetophenone	250d	20	76	98	148		See Table 27
3-Methoxyaniline (m-Anisidine)	251		80		68	138	
4-Ethoxyaniline (p-Phenetidine)	254	2	135	173	106	118	
Methyl 2-aminobenzoate (Methyl anthranilate)	255d	25	101	100			See Table 19
4-(Diethylamino)aniline	262		104	172			
Ethyl 2-aminobenzoate (Ethyl anthranilate)	265d	13	61	98	112	164	See Table 19
4-Toluidine	201	45	153 (147)	158	118	136	
3,4-Dimethylaniline		48	99	118	154	141	
2-Phenylaniline		49	118	102			
1-Naphthylamine		50	160	161	157	190	Carcinogenic
4-Phenylaniline		53	175	233	255		
4-(Dimethylamino)aniline		53	131	228			
4-Methoxyaniline (p-Anisidine)		57	127	154	114	141	
2-Aminopyridine		58	71	169			
1,3-Phenylenediamine		63	190 di / 88* mono	240 di / 125* mono	172	172	*More easily prepared than the di-amides
3-Aminopyridine		64	133	119			
2-Nitroaniline		71	93	98	115 (142)		Orange
4-Aminodiphenylamine		75	158	203		190	
4-Methyl-3-nitroaniline		78	145	172	164		Yellow
2,4-Diaminophenol		79	180 tri / 220 di-N	253 di-N			
2,5-Dimethoxyaniline		83	91	85	80	186	
4-Methyl-1,2-phenylenediamine (3,4-Diaminotoluene)		89	210	263	140 mono		
Ethyl 4-aminobenzoate (Benzocaine)		90	110	148			See Table 19
2-Methyl-6-nitroaniline		97	158	167	122		Orange
3-Aminoacetophenone		99	128		130		See Table 27
4-Methyl-1,3-phenylenediamine (2,4-Diaminotoluene)		99	224	224	192	184	

Table 9. Amines, primary aromatic (C, H, (O) and N) (*cont.*)

	B.p.	M.p.	Ethanoyl deriv. (p. 46)	Benzoyl deriv. (p. 46)	Toluene-4-sulphonyl deriv. (p. 46)	2,4-Dinitrophenyl deriv. (p. 47)	Notes
1,2-Phenylenediamine		102	186	301	202		
4-Aminoacetophenone		106	167	205	203		
2-Methyl-5-nitroaniline		107	151	183			
2-Naphthylamine		112	143	162	133	179	Carcinogenic
3-Nitroaniline		114	154	155	138	193	Yellow
4-Methyl-2-nitroaniline		117	96	148	166 (146)		Red
3-Aminophenol		122	101 di 148 mono-N	153 di 174 mono-N	157 mono		See Table 31 Mono-O-benzoyl, 153
4,4'-Diaminodiphenyl (Benzidine)		127	317 di 199 mono	352 di 203 mono	243		Carcinogenic
4,4'-Diamino-3,3'-dimethyl-diphenyl (o-Tolidine)		129	314 di 103 mono	265 di 198 mono			Picrate, 185 (p. 47) Carcinogenic
2-Methyl-4-nitroaniline		130	202	178	174		Light yellow
1,4-Phenylenediamine		140	303 di 162 mono	338 di 128 mono	266	177	
2-Aminobenzoic acid (Anthranilic acid)		144	185	182	217		See Table 17
4-Nitroaniline		147	216	199	191	186	Yellow
3,5-Dinitro-2-hydroxyaniline (2-Amino-4,6-dinitrophenol, Picramic acid)		168	201 N- 193 O-	230 N- 220 O-	191 N-		Red. See Table 31
2-Aminophenol		174d	124 di 201* mono	165 N- 185 O-	146 (139)		See Table 31 *Normal product of acetylation
3-Aminobenzoic acid		174	250	248			See Table 17
2,4-Dinitroaniline		180	120	202	219		Yellow; gives red colour with dil. NaOH and propanone(acetone)
4-Aminophenol		184d	150 di 168 N-	234 di 216 N-	168 di 252 N- 142 O-		See Table 31
4-Aminobenzoic acid		186	252	278	223		See Table 17
2,4,6-Trinitroaniline		190	230	196			Yellow

Table 10. Amines, primary aromatic (C, H, (O), N and halogen or S)

	B.p.	M.p.	Ethanoyl deriv. (p. 46)	Benzoyl deriv. (p. 46)	Toluene-4-sulphonyl deriv. (p. 46)	2,4-Dinitrophenyl deriv. (p. 47)	Notes
4-Fluoroaniline	188		152	185			
2-Chloroaniline	208		88	99	105	150	
2-Chloro-4-methylaniline	223		118	138	103		
3-Chloroaniline	230		73	122	138	184	
2-Aminothiophenol	234	26	135	154			
2-Bromo-4-methylaniline	240	26	118	149			
4-Chloro-2-methylaniline	241	29	140	172	145		
3-Chloro-2-methylaniline	245		159	173			
2-Bromoaniline	250 (229)	32	100	116	90	161	
3-Bromoaniline	251	18	88	135		178	
3-Bromo-4-methylaniline	254	25	113	132			
4-Aminothiophenol	140/16 mm di 154 N-	46	144	180 N-			
2,5-Dichloroaniline	251	50	133	120			
4-Bromo-2-methylaniline	240	59	157	115			
2,4-Dichloroaniline		63	146	117	126	116	
4-Bromoaniline		66	167	202	101	158	
4-Chloroaniline		70	178	192	95	167	
2,4,6-Trichloroaniline		78	205	174			Insol. in HCl; diazotize in ethanol and H₂SO₄
4-Chloro-1,3-phenylenediamine (2,4-Diaminochlorobenzene)		88	243* di	178	215		*Monoethanoyl deriv., 170
2-Bromo-4-nitroaniline		105	129	160			Yellow
2-Chloro-4-nitroaniline		108	139	161	164		Yellow
4-Bromo-2-nitroaniline		111	104	137			Orange
4-Chloro-2-nitroaniline		116	104		110		Orange
2,4,6-Tribromoaniline		119	232 mono 127 di	198			Insol. in HCl; diazotize in ethanol and H₂SO₄
4-Aminobenzenesulphonamide (Sulphanilamide)		165	219 4-N- 254 di	284 4-N- 268 di			See Table 33

Table 11. Amines, secondary

	B.p.	M.p.	Ethanoyl deriv. (p. 46)	Benzoyl deriv. (p. 46)	Toluene-4-sulphonyl deriv. (p. 46)	2,4-Dinitrophenyl deriv. (p. 47)	Notes
Dimethylamine	7			42	79	87	Normally in aq. solution
Diethylamine	55			42	60	80	
Di-isopropylamine	84						Picrate, 140 (p. 47)
Pyrrolidine	89				123		Picrate, 112 (p. 47)
Piperidine	106			48	103	93	Miscible with water
Dipropylamine	110					40	Picrate, 75 (p. 47)
2-Methylpiperidine	117			45	55		Picrate, 135 (p. 47)
3-Methylpiperidine	124						Picrate, 137 (p. 47)
Morpholine	130			75	147		
Di-isobutylamine	139		86			112	
N-Methylcyclohexyl-amine	146			85			Picrate, 170 (p. 47)
2-Ethylpiperidine	146						Picrate, 133 (p. 47)
Dibutylamine	159						Picrate 60 (p. 47)
N-Methylaniline	194		101	63	95	167	Picrate, 145
N-Ethylaniline	205		54	60	88	95	
N-Methyl-3-toluidine	206		66				
N-Methyl-2-toluidine	207		56	66	120	155	
N-Methyl-4-toluidine	208		83	53	60		
N-Ethyl-2-toluidine	214			72	75	114	
N-Ethyl-4-toluidine	217			40	71	120	
N-Ethyl-3-toluidine	221			72			Picrate, 132 (p. 47)
1,2,3,4-Tetrahydro isoquinoline	233		46	129			Picrate, 200 (p. 47)
1,2,3,4-Tetrahydro-quinoline	250	20		75			Picrate, 141 (p. 47)
Dicyclohexylamine	254	20	103	153	119		
Di-(2-hydroxyethyl)amine (Diethanolamine)	270	28			99		Picrate, 110 (p. 47); see Table 3
Dibenzylamine	300			112	158	105	
Piperazine hexahydrate		44*	134 di 52 mono	191 di 75 mono	173 mono		*Anhydrous, 104, and is sol. in water but only slightly in ether
Indole		52		68			Picrate, 187
Diphenylamine		54	101	180 (107*)	142		*Resolidifies at 135 and melts at 180
N-Phenyl-1-naphthylamine		62	115	152			
N-Phenyl-2-naphthylamine		108	93	148 (111)			
Carbazole		243	69	98	137		Very weakly basic

Table 12. Amines, tertiary

	B.p.	M.p.	Meth-iodide (p. 47)	Picrate (p. 47)	Notes
Trimethylamine	3		230	216	
Triethylamine	89		280	173	
Pyridine	116		117	167	
2-Methylpyridine	129		230	169	
(α-Picoline)					
2-(Dimethylamino)ethanol	135			96	
2,6-Dimethylpyridine	143		238	161	
(2,6-Lutidine)					
3-Methylpyridine	144		92	150	
(β-Picoline)					
4-Methylpyridine	144		152	167	
(γ-Picoline)					
2-Ethylpyridine	149			187	
Tripropylamine	156		208	117	
2,4-Dimethylpyridine	158		113	180	
(2,4-Lutidine)					
2-(Diethylamino)ethanol	161		249d	79	
4-Ethylpyridine	164			168	
2-Chloropyridine	166				Methyl toluene-4-sulphonate salt, 120 (p. 47)
3,5-Dimethylpyridine	170			238	
(3,5-Lutidine)					
2,4,6-Trimethylpyridine	172			156	
(2,4,6-Collidine)					
NN-Dimethyl-2-toluidine	185		210	122	
NN-Dimethylaniline	193		218*	162	*Sublimes. 4-Nitroso deriv., 87 (p. 47)
N-Ethyl-N-methylaniline	201		125	134	4-Nitroso deriv., 66 (p. 47)
NN-Dimethyl-4-toluidine	210		215*	130	*Sublimes
Tributylamine	211		180	107	
NN-Dimethyl-3-toluidine	212		177	131	
NN-Diethylaniline	216		104	142	4-Nitroso deriv., 84 (p. 47)
Quinoline	238		72*	203	*Hydrate, anhydrous, 133
Isoquinoline	243	24	159	222	
NN-Dipropylaniline	245		156	261	
2-Methylquinoline	247		195	191	
(Quinaldine)					
8-Methylquinoline	248		193	203	
6-Methylquinoline	258		219	234	
4-Methylquinoline (Lepidine)	262		174	212	
2,4-Dimethylquinoline	264		252	193	
4-(Diethylamino)benzaldehyde		41			Yellow; see Table 5
4-Bromo-NN-dimethylaniline		55	185		
2,6-Dimethylquinoline		60	244d	191	
4-(Dimethylamino)benzaldehyde		74			See Table 5
8-Hydroxyquinoline		75	143	204	See also Table 31
Tribenzylamine		92	184	190	
2,3-Dimethyl-1-phenylpyrazol-5-one		111		181	4-Nitroso deriv., 200 Absorbs bromine; gives orange colour with aq. FeCl₃
(Phenazone, antipyrine)					
Triphenylamine		127			Not basic; nitration in ethanoic acid gives trinitro deriv., 280
4,4'-Di-(dimethylamino)benzophenone		174	105	156	Oxime, 233 (p. 45). See Table 27
(Michler's ketone)					
Hexamethylene tetramine		280*	190	179	*Sublimes

77

Table 13. Amino-acids

	Decomposition temp.	Benzoyl deriv. (p. 48)	3,5-Dinitrobenzoyl deriv. (p. 48)	Toluene-4-sulphonyl deriv. (p. 48)	Notes
N-Phenylglycine	126	63			Ethanoyl deriv., 194 (p. 48)
(+)- or (−)-Orithine	140*	188 di 240 mono			*Often a syrup. Picrate, 204 (p. 48)
2-Aminobenzoic acid (Anthranilic acid)	144	182	278		Ethanoyl deriv., 185 (p. 48); see Tables 9 and 17
3-Aminobenzoic acid	174	248			Ethanoyl deriv., 250 (p. 48); see Tables 9 and 17
4-Aminobenzoic acid	186	278			Ethanoyl deriv., 252 (p. 48); see Tables 9 and 17
3-Aminopropanoic acid (β-Alanine)	196		202	117	
Glutamic acid	199	157		117	Ethanoyl deriv., 185 (p. 48)
4-Aminophenylacetic acid	200	205			Ethanoyl deriv., 170 (p. 48)
Proline	203*		217		*Monohydrate, 190. Picrate, 135
(+)- or (−)-Arginine	207	298 mono 235 di	150		Picrate, 206 (p. 48)
(+)- or (−)-Glutamic acid	211	138	217	131	
Sarcosine	212	103	154	102	
(+)- or (−)-Proline	222	156		133	Picrate, 154 (p. 48)
(+)- or (−)-Lysine	224	149	169		Picrate, 266 (p. 48)
(+)- or (−)-Asparagine	226	189	196	175	
Serine	228	171 mono 124 di	95	213	
Glycine	232	187	179	147	Ethanoyl deriv., 206 (p. 48)
Threonine	235	174* di 176* mono			*Mixed m.p. 145
Arginine	238	230*			*Hydrate, 176. Picrate, 201 (mono), 196 (di) (p. 48)
(+)- or (−)-Cystine	260	181	180	201	
(+)- or (−)-Aspartic acid	271	185		140	
Methionine	281 (272)	151		105	Ethanoyl deriv., 114 (p. 48)
Phenylalanine	273	188	93	134	
Tryptophan	275	188	240	176	
(+)- or (−)-Histidine	277	249	189	203	
2-Amino-2-methylpropionic acid	280*	198			*Sublimes
Aspartic acid	280	165			
(+)- or (−)-Methionine	283	150	95		Ethanoyl deriv., 98 (p. 48)
(+)- or (−)-Tryptophan	289	176	233d	176	
Isoleucine	292	118		140	
Alanine	295	166	177	139	
(+)- or (−)-Alanine	297	151		133	
Valine	298	132	158	110	
2-Aminobutanoic acid	307	147	194		

Table 13. Amino-acids (*cont.*)

	Decomposition temp.	Benzoyl deriv. (p. 48)	3,5-Dinitrobenzoyl deriv. (p. 48)	Toluene-4-sulphonyl deriv. (p. 48)	Notes
(+)- or (−)-Valine	315	127	158	147	
Tyrosine	318	197 mono-*N*	254	224	
(+)- or (−)-Phenylalanine	320	146	93	164	
Norleucine	327			124	
Leucine	332	141	187		Ethanoyl deriv., 157
(+)- or (−)-Leucine	337	107*	187	124	*Hydrate, 60
(+)- or (−)-Tyrosine	344	166 mono-*N* 211 di		188 mono-*N* 119 di	Ethanoyl deriv., 172 (p. 48)
Ornithine		267d mono 188 di		188 mono	
Lysine		249 mono 145 di			Picrate, 225d (mono) (p. 48)

Table 14. Azo, azoxy, nitroso and hydrazine compounds

	B.P.	Notes
NN-Dimethylhydrazine	63	Picrate, 146 (p. 47)
NN'-Dimethylhydrazine	81	Picrate, 148 (p. 47)
Phenylhydrazine (m.p. 19)	243	Benzoyl deriv., 168; with benzaldehyde → hydrazone, 158
	M.p.	
NN-Diphenylhydrazine	34	Benzoyl deriv., 192; ethanoyl deriv., 184; with benzaldehyde→hydrazone, 122
Azoxybenzene	36	Light yellow. Warming with conc. H_2SO_4 → 4-hydroxyazobenzene, 152
4-Tolylhydrazine	66	Benzoyl deriv., 146; ethanoyl deriv., 131; with benzaldehyde → hydrazone, 125
Nitrosobenzene	68	Colourless solid but turns green on melting. With Sn + HCl → aniline (p. 58)
Azobenzene	68	Orange-red colour. Zn dust + ethanolic NaOH → hydrazobenzene, 131
NN-Diethyl-4-nitrosoaniline	84	Dark green, forms yellow hydrochloride; $KMnO_4$ → *NN*-diethyl-4-nitroaniline, 77
NN-Dimethyl-4-nitrosoaniline	85	Dark green, forms yellow hydrochloride; $KMnO_4$ → *NN*-dimethyl-4-nitroaniline, 163
1-Nitroso-2-naphthol	109	Orange. Benzoyl deriv., 114. Cold dil. HNO_3 → 1 —NO_2 deriv., 103
4-Nitrosophenol	125d	Colourless. Ethanoic anhydride at 100° → ethanoyl deriv., 107 (yellow)
NN'-Diphenylhydrazine (Hydrazobenzene)	127	Dibenzoyl deriv., 162 (p. 43)
4-Nitrosodiphenylamine	145	Green. Zn + ethanoic acid → *p*-amino deriv., 66
1,2-Di-(2-hydroxyphenyl)hydrazine (2,2'-Hydrazophenol)	148	Yellow. Benzoyl deriv., 186; see also Table 31
4-Hydroxyazobenzene	152	Yellow. Ethanoyl deriv., 84; benzoyl deriv., 138 (p. 43)
4-Nitrophenylhydrazine	157d	Orange. Picrate, 119; benzaldehyde → hydrazone, 192
2,4-Dinitrophenylhydrazine	197d	Red. Ethanoyl deriv., 197; benzoyl deriv., 206

Table 15. Carbohydrates

	Approximate decomposition temperature	[α]$_D$ (in water) Initial	[α]$_D$ (in water) Final	4-N-Glycosylaminobenzoic acid (p. 49)	Ethanoate* (p. 48)	Osazone (p. 49)	Notes
Melibiose, monohydrate	85	+111	+129		α147 β177	178	Disaccharide
D-Ribose	90	−21	−21	156		164	
D-Glucose, monohydrate	90	+112	+52	134	α112 β132	205d	Pentabenzoate, 179 (p. 49)
Maltose, monohydrate	101	+111	+130		α125 β160	206	Disaccharide
D-Fructose	104	−132	−92		α 70 β109	205d	Pentabenzoate, 79 (p. 49)
L-Rhamnose, monohydrate	105	−8.6	+8.2	170	99	182	
L-Rhamnose, anhydrous	123	−8.6	+8.2	170	99	182	
D-Mannose	132	+29	+14	182	α 74 β115	205d	
D-Xylose	144	+93	+18	181	α 59 β126	160	
D-Glucose, anhydrous	146	+112	+52	134	α112 β132	205d	
D- or L-Arabinose	160	−175 D- +190 L-	−105 D- +104 L-	192	α 94 β 86	166	
L-Sorbose	161	−43	−43		97	162	
Maltose, anhydrous	165	+111	+130		α125 β160	206	Disaccharide
D-Galactose	170	+150	+80	160	α 95 β142	196	
Sucrose	185	+66	+66		70	205	Non-reducing disaccharide
Lactose, monohydrate	203	+90	+55		α152 β 90	200d	Disaccharide
Cellobiose	225	+14	+35		α229 β192	200	Disaccharide

*The preparation of the β-form is described on p. 48 but the melting points of both anomers are given because a small amount of the α-form is sometimes formed.

Table: Carboxylic acids (C, H and O), their acyl chlorides, anhydrides and amines

Name	B.p.	M.p.	Chloride B.p.	Anhydride B.p.	Nitrile B.p.	Amide (p. 49) M.p.	Anilide (p. 49) M.p.	4-Toluidide (p. 49) M.p.	4-Bromophenacyl ester (p. 50) M.p.	4-Phenylphenacyl ester (p. 50) M.p.	Notes
Methanoic (Formic)	100	8	—	—	26	3	50*	53*	140†	74	*By heating the acid with the amine †ArCO·CH₂·OH is often isolated; ester melts at 99
Ethanoic (Acetic)	118	16	52	140	82	82	114	153 (147)	85	111	
Propanoic	140		80	168	97	81	106	126	61	102	
Propenoic (Acrylic)	140	13	75		78	84	104	141			Unsaturated; polymerizes readily
Propynoic (Propiolic)	144d	18				61	87				Unsaturated
2-Methylpropanoic (Isobutyric)	155		92	182	108	128	105	108	77	89	
2-Methylpropenoic (Methacrylic)	161	16	95		90	102	87				Unsaturated
Butanoic (Butyric)	163		101	198	118	116	96	75	63	97	
2,2-Dimethylpropanoic (Pivalic)	164	35	105	190	106	155	132	119	76	114	
2-Oxopropanoic (Pyruvic)	165	13				124	104	109			See Table 26
But-3-enoic (Vinylacetic)	169 (163)		98			73	58		60		Unsaturated
cis-But-2-enoic (Isocrotonic)	169	15				102	101	132	81	71	Unsaturated
2-Methylbutanoic	177		115	215	125	112	110	93	55	78	
3-Methylbutanoic (Isovaleric)	177		115	218	129	135	110	107	68	64	
Pentanoic (Valeric)	186		126	229	140	106	63	74	75	77	
2-Ethylbutanoic	195		139		145	112	127	116	77	70	
4-Methylpentanoic (Isocaproic)	199				155	120	112	63	72	68	
Hexanoic (Caproic)	205		153	254	163	100	94	74	72	62	
Heptanoic	223		193	258	184	96	65	81		53	
2-Ethylhexanoic	227		88/20 mm	150/8 mm	75/9 mm	102	89	107	66	67	
Octanoic (Caprylic)	237	16	195	281	205	106	57	70	84		See Table 26
4-Oxopentanoic (Laevulinic, laevulic)	245d	33				107	102	109	69	71	
Nonanoic	254	12	215		224	101	57	84			
2-Phenylpropanoic	265		97/12 mm		230	95			66		
Decanoic (Capric)	269	31	114/15 mm	24*	245	108	70	78			*M.p.
Undec-10-enoic (Undecylenic)	275	24				87	67	68	68	80	Unsaturated
Undecanoic	280	28		37*	248	103	71	80			*M.p.

Table 16. Carboxylic acids (C, H and O), their acyl chlorides, anhydrides and nitriles (cont.)

	B.p.	M.p.	Chloride B.p.	Anhydride B.p.	Nitrile B.p.	Amide (p. 49) M.p.	Anilide (p. 49) M.p.	4-Toluidide (p. 49) M.p.	4-Bromophenacyl ester (p. 50) M.p.	4-Phenylphenacyl ester (p. 50) M.p.	Notes
cis-Octadec-9-enoic (Oleic)	286d	16			330d	76	41	42	46	61	*M.p. Unsaturated
2-Hydroxypropanoic (Lactic)	122/15 mm	18		22*	182	78	58	107	113	145	
Dodecanoic (Lauric)		44	145/18 mm	42*	280	100	78	87	76	84	*M.p.
3-Phenylpropanoic (Hydrocinnamic)		48	225d		261	105 (82)	96	135	104	95	
trans-Octadec-9-enoic (Elaidic, *trans*-Oleic)		51		51*		93			65	73	*M.p. Unsaturated
4-Phenylbutanoic (4-Phenylbutyric)		52	119/9 mm			84			58	90	
Tetradecanoic (Myristic)		54	174/16 mm	54*	19*	102	84	93	81	90	*M.p.
Hexadecanoic (Palmitic)		62	12*	63*	31*	106	89	98	84	94	*M.p.
3-Methylbut-2-enoic (*ββ*-Dimethylacrylic)		68	145		140	107			104	145	Unsaturated
Octadecanoic (Stearic)		70	22*	70*	43*	108	94	102	90	97	*M.p.
But-2-enoic (Crotonic)		72	126	246	118	158	118	132	96		Unsaturated
Phenylethanoic (Phenylacetic)		76	210	72*	232	157	118	136	89	63d	*M.p.
2-Hydroxy-2-methylpropanoic (*α*-Hydroxyisobutyric)		79				98	136	133			
Hydroxyethanoic (Glycollic)		80		128*	183	120	97	143	138	109	Is often a syrup. *M.p.
Methylmaleic (Citraconic)		92d	95/18 mm	214		185	176 di, 153 mono	170 mono			
Pentanedioic (Glutaric)		98	218	56*	286	175	224	218	137	152	*M.p.
Phenoxyethanoic		99	225	67*	239	101	99	189	149	146	*M.p.
2-Carboxy-2-hydroxypentanedioic, monohydrate (Citric, monohydrate)		100				210	199		148	166	Heat at 130° → anhyd. acid, 153
Ethanedioic, dihydrate (Oxalic, dihydrate)		100	64			419 di, 219 mono	246 di, 148 mono	268 di, 169 mono	242		Heat → anhydrous acid, 189
(−)-Hydroxysuccinic (Malic)		100				156	197		179	106	
2-Methoxybenzoic (*o*-Anisic)		100	254		24*	129	78	207	113	131	*M.p.

	M.p.	Chloride B.p.	Anhydride B.p.	Nitrile B.p.	Amide (p. 49) M.p.	Anilide (p. 49) M.p.	4-Toluidide (p. 49) M.p.	4-Bromophenacyl ester (p. 50) M.p.	4-Phenylphenacyl ester (p. 50) M.p.	Notes
Heptanedioic (Pimelic)	104				175	156 di / 108 mono	206	137	146d	
2-Toluic	105	212	39*	205	142	125	144	57	95	*M.p.
Nonanedioic (Azelaic)	106	166/18 mm			175 di / 94 mono	187 di / 107 mono	200	131	141	*M.p.
3-Toluic	111	218	71*	212	96	126	118	108	137	
3-Benzoylpropanoic	116				125	150		98		See Table 26
3-Hydroxy-2-phenylpropanoic (Tropic)	117				169					Ethanoate, 88 (p. 43)
2-Hydroxy-2-phenylethanoic (Mandelic)	118			21*	133	151	172	113		*M.p.
Benzoic	122	197	42*	191	128	163	158	119	167	*M.p.
2-Benzoylbenzoic	128*	70†		83†	165	195				Monohydrate, 91. See Table 26. †M.p.
cis-Butenedioic (Maleic)	130 (139)	60*	56*	31*	266 di / 181 mono	187	142	168	168	*M.p. Unsaturated
Propanedioic (Malonic)	133d			219	170 di / 50 mono	225	252		175	
Decanedioic (Sebacic)	133	182/16 mm			210 di / 170 mono	200 di / 122 mono	201	147	140	
3-Phenylpropenoic (Cinnamic)	133	35*	136*	255 / 20*	147	151	168	146	183	Unsaturated
Furoic	133	173	73*	146	142	124	108	139	91	*M.p.
1-Naphthylethanoic (1-Naphthylacetic)	133	188/23 mm		183	181	155		112		*M.p.
Hexa-2,4-dienoic (Sorbic)	134	78/15 mm			168	153		129	141	Unsaturated
2-(Ethanoyloxy)benzoic (Acetylsalicylic, aspirin)	135		43*		138	136				*M.p.
3-Phenylpropynoic (Phenylpropiolic)	137	116/17 mm		41*	100	126	142			Unsaturated. *M.p.

Table 16. Carboxylic acids (C, H and O), their acyl chlorides, anhydrides and nitriles (*cont.*)

	M.p.	Chloride B.p.	Anhydride B.p.	Nitrile B.p.	Amide (p. 49) M.p.	Anilide (p. 49) M.p.	4-Toluidide (p. 49) M.p.	4-Bromophenacyl ester (p. 50) M.p.	4-Phenylphenacyl ester (p. 50) M.p.	Notes
meso-Dihydroxybutanedioic (*meso*-Tartaric)	140			131*	190	193 mono				*M.p.
Octanedioic (Suberic)	142	162/15 mm	65*		216 di 125 mono	187 di 128 mono	219	144	151	*M.p.
3,4,5-Trimethoxybenzoic	144				184	154		129		
Diphenylethanoic	148	56*	98*	72*	168	180	172	112	111	*M.p.
Benzilic (Diphenylglycollic)	150	193/27 mm			154	175	189	152	122	
3-Carboxy-3-hydroxypentanedioic, anhydrous (Citric)	153				210*	199*	189*	148	146	*From amine and ester on prolonged heating
Hexanedioic (Adipic)	153	130/18 mm		295	220 di 126 mono	238 di 151 mono	241	155	148	
2-Hydroxy-5-methylbenzoic (5-Methylsalicylic)	153				177			142		See Table 30
2-Hydroxybenzoic (Salicylic)	158			98*	139	135	156	140	148	See Table 30. *M.p.; 5-NO$_2$ deriv. (use method (i) p. 52), 228
1-Naphthoic	161	20*	145*	35*	202	163		135	126	*M.p.
2-Hydroxy-3-methylbenzoic (3-Methylsalicylic)	163				112		164			See Table 30
Methylenebutanedioic (Itaconic)	165				192	190* mono		117		*By heating acid with an excess of amine. Unsaturated
4-*t*-Butylbenzoic	165				173					
Phenylbutanedioic	167				211	222				
(+)-Dihydroxybutanedioic ((+)-Tartaric)	169				195 di 171 mono	264 di 180 mono		216	204d	Dibenzoate, 90
Butynedioic (Acetylenedicarboxylic)	179			218	294d					Unsaturated
4-Toluic	180	214	95*		160	144	160	153	165	*M.p.
3,4-Dimethoxybenzoic (Veratric)	181				164	154		124		
4-Methoxybenzoic (*p*-Anisic)	184	22*	99*	61*	163	169	186	152	160	*M.p.

84

Acid	m.p.								*M.p.
2-Carboxyphenylethanoic (Homophthalic)	185		141*	228	232				*M.p. Ethanoate, 158. See Table 30
2-Naphthoic	185	43*	133*	192	171	191	211	183	*M.p. See Table 30
Butanedioic (Succinic)	185	20*	119*	260 di / 157 mono	230 di / 148 mono	255 di / 179 mono	211	208	*M.p. See Table 30
(+)-Camphoric	187			192 di / 177 mono	226 di / 204 mono				Loses water near m.p. to form anhydride
Ethanedioic, anhydrous (Oxalic)	188	64		419 di / 219 mono / 202 mono	246 di / 148 mono / 154 mono	268 di / 169 mono	242	166	*M.p. †Loses NH$_3$ near m.p. to form imide, 233
1-Hydroxy-2-naphthoic	195	85*	82*						Dibenzoate, 112
3-Hydroxybenzoic	200			170	156	163	176 (168)		*M.p. †Difficult to prepare See Table 30
3,4-Dihydroxybenzoic (Protocatechuic)	200d	98*	156*	212	166				*M.p. See Table 30
2,5-Dihydroxybenzoic (Gentisic)	200			218					*M.p. See Table 30
Benzene-1,2-dicarboxylic (Phthalic)	200d	275	131*	220†	254 di / 170 mono	201 di / 155 mono	153	167	
Dihydroxybutanedioic (Tartaric)	205		113*	226	235	204†	191	240	*M.p. †Difficult to prepare See Table 30
4-Hydroxybenzoic	213		162†	196†		204†			*M.p. See Table 30
2,4-Dihydroxybenzoic (β-Resorcylic)	213d		175*	222	126				See Table 30
Galactaric (Mucic)	214d			222 / 220 di / 192 mono			225	149d	*M.p. Tetra-ethanoate, 266 (p. 43)
3-Hydroxy-2-naphthoic	222	95*	188*	217	243	221			*M.p. Ethanoate, 184. See Table 30
3,5-Dihydroxybenzoic (α-Resorcylic)	233			245* (189)	207*				See Table 30
3,4,5-Trihydroxybenzoic (Gallic)	240d			267	314		134	198d	*Difficult to prepare. See Table 30
trans-Butenedioic (Fumaric)	287*	160	96†	267				256d	*In a sealed tube; sublimes at 200. Unsaturated †M.p.
Benzene-1,4-dicarboxylic (Terephthalic)	>300*	83†	222†	>350d	334	225	225		*Sublimes. †M.p.
Benzene-1,3-dicarboxylic (Isophthalic)	345	41*	162*	280	250	179		280	*M.p.

Table 17. Carboxylic acids (C, H, O and halogen, N or S)

	B.p.	M.p.	Amide (p. 49)	Anilide (p. 49)	4-Toluidide (p. 49)	4-Bromo-phenacyl ester (p. 50)	4-Phenyl phenacyl ester (p. 50)	Notes
Trifluoroethanoic (Trifluoroacetic)	72		74	91				Acid chloride b.p. −27
Thioacetic	93	31	108					Pale yellow; unpleasant odour
Fluoroethanoic	167		108	76	130			
2-Chloropropanoic	186		80	92	124			
Dichloroethanoic	194	5	98	125	153	99		
2-Bromopropanoic	203	25	123	99	125			
2-Bromobutanoic (2-Bromobutyric)	217d		112	98	92			
Mercaptoethanoic (Thioglycollic)	123/29 mm		52	111	125			
3-Chloropropanoic		40	101	116	124			
2-Bromo-3-methylbutanoic		44	133	83	93			
2-Bromo-2-methylpropanoic		48	148		91			
Bromoethanoic	208	50	91	131	113			
Trichloroethanoic	196	57	141	94				
Chloroethanoic		63	120	137	162	105	116	
Cyanoethanoic		66	123	198				
Iodoethanoic		83	95	143				
2-Hydroxy-3-nitrobenzoic (3-Nitrosalicylic) hydrate		125	145		165			
Pyridine-2-carboxylic (2-Picolinic)		138	107	76	104			
2-Chloro-4-nitrobenzoic		139	172	168				
3-Nitrobenzoic		140	142	154	162	137	153	
2-Nitrophenylethanoic		141	161					
4-Chloro-2-nitrobenzoic		142	172	131				
2-Chlorobenzoic		142	142	118	131	106	123	
2-Hydroxy-3-nitrobenzoic, anhyd.		144	145					

Acid							Notes
2-Aminobenzoic (Anthranilic)	144	109	131	151	172*		*Two moles of reagent are required to form this derivative. See Tables 9 and 13
2-Nitrobenzoic	146	175	155	203	107	140	
2-Bromobenzoic	150	155	141		102	98	
4-Nitrophenylethanoic	152	198	198	210	207		
3-Bromobenzoic	155	155	146		126	155	
3-Chlorobenzoic	158	134	124		117	154	
2-Iodobenzoic	162	184	142		110	143	
2,4-Dichlorobenzoic	164	194	165 (153)	168			Conc. HNO_3 + $H_2SO_4 \rightarrow$ 3,5-dinitro deriv., 212
5-Bromo-2-hydroxybenzoic	165	232	222				Ethanoate, 168 (p. 59)
4-Nitrophthalic	165	200d	192	172		120	
2-Mercaptobenzoic	165						Ethanoate, 125 (p. 59)
3-Aminobenzoic	174	111	140		190*		See Tables 9 and 13. *Two moles of reagent are required to form this derivative
4-Chloro-3-nitrobenzoic	181	156	131				
4-Fluorobenzoic	182	154					
2,4-Dinitrobenzoic	183	203			158		
4-Aminobenzoic	186	114			200*		See Tables 9 and 13. *See under 3-aminobenzoic acid
3-Iodobenzoic	187	186			128	147	
N-Benzoylglycine (Hippuric)	187	183			151	163	
3,5-Dinitrobenzoic	204	183	208	280	159	154	
3-Nitrophthalic	218	201d	235	224	166	149	Acyl chloride, 70 Anhydride, 162
2-Hydroxy-5-nitrobenzoic	229	225	234				
Pyridine-3-carboxylic (Nicotinic)	237	122	224	150	142	146	
3-(2-Nitrophenyl)propenoic (2-Nitrocinnamic)	239	185	132*				*From benzene; from water, 85 Unsaturated
4-Nitrobenzoic	240	201	211	203	134	182	Acyl chloride, 75
4-Chlorobenzoic	241	179	194		126	160	
4-Bromobenzoic	251	189	197		134	160	
4-Iodobenzoic	265	217	210		146	171	
3-(4-Nitrophenyl)propenoic (4-Nitrocinnamic)	285	217			191	192	Unsaturated
Pyridine-4-carboxylic (Isonicotinic)	324*	155					*Sublimes

Table 18. Enols

	B.p.	Semi-carbazone (p. 50)	2,4-Dinitro-phenyl-hydrazone (p. 50)	Colour with aq. FeCl$_3$	Notes
Pentan-2,4-dione (Acetylacetone)	139	107*	122* 209 di	Red	*Pyrazole deriv. See Table 26
Methyl 3-oxobutanoate (Methyl acetoacetate)	170	152	119	Red	See Table 26
Ethyl 3-oxobutanoate (Ethyl acetoacetate)	181d	133	96	Red	See Table 26
Ethyl 3-oxopentanedioate (Ethyl acetonedicarboxylate)	250	94	86	Red	See Table 26
	M.p.				
1-Phenylbutan-1,3-dione (Benzoylacetone)	61		151	Red	See Table 26
1,3,5-Trihydroxybenzene (Phloroglucinol)	217			Violet*	*Transient colour. Picrate, 101. See Table 30

Table 19. Esters, carboxylic

	B.p.	Notes
Methyl methanoate	32	
Ethyl methanoate	54	
Methyl ethanoate	57	
Isopropyl methanoate	68	
Vinyl ethanoate	72	Unsaturated; polymerizes readily
Methyl chloromethanoate	73	Very reactive chlorine atom
Ethyl ethanoate	77	
Methyl propanoate	79	
Propyl methanoate	81	
Prop-2-enyl methanoate	83	Unsaturated
Methyl propenoate	85	Unsaturated; polymerizes readily
Isopropyl ethanoate	91	
Methyl 2-methylpropanoate	92	
Ethyl chloromethanoate	93	Very reactive chlorine atom
2-Methylpropyl methanoate	98	
Ethyl propanoate	98	
2-Methylprop-2-yl ethanoate	98	
Methyl 2-methylpropenoate	99	Unsaturated; polymerizes readily
Propyl ethanoate	101	
Ethyl propenoate	101	Unsaturated; polymerizes readily
Methyl butanoate	102	
Prop-2-enyl ethanoate	103	Unsaturated
Butyl methanoate	107	
Ethyl 2-methylpropanoate	110	
But-2-yl ethanoate	111	
Methyl 3-methylbutanoate	116	
Ethyl butanoate	120	
3-Methylbutyl methanoate	123	
Butyl ethanoate	125	
Methyl pentanoate	130	
2-Methylpropyl chloromethanoate	130	Very reactive chlorine atom
Methyl chloroethanoate	130	Very reactive chlorine atom

Table 19. Esters, carboxylic (*cont.*)

	B.p.	Notes
Methyl 2-oxopropanoate	136	2,4-Dinitrophenylhydrazone, 187 (p. 44). See Table 26
Ethyl but-2-enoate	138	Unsaturated
3-Methylbutyl ethanoate	142	
Methyl bromoethanoate	144d	Very reactive bromine atom
Ethyl chloroethanoate	145	Very reactive chlorine atom
Ethyl 2-chloropropanoate	146	Very reactive chlorine atom
Methyl hexanoate	150	
Ethyl 2-oxopropanoate	155	Semicarbazone, 206d (p. 44); 2,4-dinitrophenyl-hydrazone, 155 (p. 44)
Ethyl bromoethanoate	159	Very reactive bromine atom
Ethyl 2-bromopropanoate	162	Very reactive bromine atom
Ethyl hexanoate	166	
Ethyl trichloroethanoate	167	
Methyl 3-oxobutanoate	170	Red colour with aq. $FeCl_3$; semicarbazone, 152 (p. 44); see Table 26
Methyl heptanoate	173	
Ethyl 3-bromopropanoate	179	
Methyl 2-furoate	181	
Dimethyl propanedioate	181	
Ethyl 3-oxobutanoate	181	Red colour with aq. $FeCl_3$; semicarbazone, 129 (p. 44); see Table 26
Diethyl ethanedioate	186	
Methyl octanoate	193	
Dimethyl butanedioate	195	
Methyl 4-oxopentanoate	196	See Table 26
Phenyl ethanoate	196	
Methyl benzoate	198	
Diethyl propanedioate	199	
Methyl cyanoethanoate	200	
γ-Butyrolactone	204	
Ethyl 4-oxopentanoate	205	Semicarbazone, 135 (p. 44)
Dimethyl *cis*-butenedioate	205	Unsaturated
Ethyl octanoate	206	
γ-Pentanolactone	207	
Ethyl benzoate	212	
Benzyl ethanoate	214	
Diethyl *trans*-butenedioate	216	Unsaturated
Diethyl butanedioate	216	
Methyl phenylethanoate	220	
Methyl 2-hydroxybenzoate	224	See Table 30
Diethyl *cis*-butenedioate	225	Unsaturated
(−)-Menthyl ethanoate	227	
Ethyl phenylethanoate	227	
Ethyl 2-hydroxybenzoate	233	See Table 30
Diethyl hexanedioate	245	
Methyl undecyl-10-enoate	248	Unsaturated
Diethyl 3-oxopentanedioate	250	Red colour with aq. $FeCl_3$; semicarbazone, 94 (p. 44); see Table 26
Methyl 2-aminobenzoate	255d	See Table 9
Glyceryl triethanoate	258	
Glyceryl monoethanoate	260	
Methyl 3-phenylpropenoate (m.p. 33)	263	Unsaturated
Ethyl 2-aminobenzoate	265d	See Table 9
Ethyl benzoylethanoate	265	See Tables 18 and 26
Ethyl 3-phenylpropenoate (m.p. 12)	271	Unsaturated
Diethyl dihydroxybutanedioate (m.p. 17)	280	
Dimethyl benzene-1,2-dicarboxylate	282	

Table 19. Esters, carboxylic (*cont.*)

	B.p.	Notes
Diethyl benzene-1,2-dicarboxylate	298	
Diethyl benzene-1,3-dicarboxylate	302	
Benzyl benzoate	323	
Dibutyl benzene-1,2-dicarboxylate	338	
Methyl octadecanoate (m.p. 38)	214/15 mm	

	M.p.	
Glyceryl 1,3-diethanoate	40	
Dibenzyl benzene-1,2-dicarboxylate	42	
Phenyl 2-hydroxybenzoate	42	See Table 30
Diethyl benzene-1,4-dicarboxylate	43	
(+)-Dimethyl dihydroxybutanedioate	48	
1-Naphthyl ethanoate	49	
Dimethyl ethanedioate	52	
Ethyl 4-nitrobenzoate	56	
Dimethyl benzene-1,3-dicarboxylate	67	
Phenyl benzoate	68	
Methyl 3-hydroxybenzoate	70	Violet colour with aq. FeCl$_3$
2-Naphthyl ethanoate	70	
Glyceryl tristearate	71	
Methyl 3-nitrobenzoate	78	
Ethyl 4-aminobenzoate	90	See Table 9
Methyl 4-nitrobenzoate	96	
Dimethyl *trans*-butenedioate	102	Unsaturated
Ethyl 4-hydroxybenzoate	115	Violet colour with aq. FeCl$_3$. See Table 30. Benzoate 89 (p. 59)
Methyl 4-hydroxybenzoate	131	Violet colour with aq. FeCl$_3$. See Table 30. Benzoate, 135. (p. 59)
Dimethyl benzene-1,4-dicarboxylate	140	

Table 20. Esters, phosphoric. Hydrolysis of the ester (p. 51) gives an alcohol or phenol which should be identified in the usual way

	B.p.
Trimethyl phosphate	197
Triethyl phosphate	215
Tripropyl phosphate	138/47 mm
Tri-2-tolyl phosphate	265/20 mm
Tributyl phosphate	157/10 mm

	M.p.
Triphenyl phosphate	50
Tribenzyl phosphate	64
Tri-4-tolyl phosphate	78
Tri-2-phenylphenyl phosphate	113

Table 21. Ethers

	B.p.	M.p.	Alkyl 3,5-dinitrobenzoate (p. 52)	Picric acid complex (p. 53)	Sulphonamide (p. 53) position	Sulphonamide M.p.	Nitro deriv. (p. 52) position	Nitro deriv. M.p.	Nitro deriv. method	Notes
Diethyl ether	35		93							
Chloromethyl methyl ether	59			163						
Tetrahydrofuran	65									
Di-isopropyl ether	68		122							
Dipropyl ether	90		75							
Di-(prop-2-enyl) ether	94		50							Unsaturated
Dioxan	101	11								Miscible with water
1-Chloro-2,3-epoxypropane (Epichlorohydrin)	116									Boiling ethanolic NaOH + phenol → glycerol diphenyl ether 81
Di-(1-methylpropyl) ether	121		76							
Di-(2-methylpropyl) ether	122		88							
2-Ethoxyethanol (Ethylene glycol monoethyl ether)	135		75							Miscible with water
Dibutyl ether	142		64							
Methoxybenzene (Anisole)	154				4	111	2,4	94* 87*	i	*Two crystalline forms
2-Methoxytoluene	171		69	117	5	137	3,5	69	iv	
Ethoxybenzene (Phenetole)	172			92	4	149	4	59	iii	
Benzyl methyl ether	172			116						
4-Methoxytoluene	176			89	3	182	3,5	122	ii	
3-Methoxytoluene	177			114	6	130	2	54	v	
							2,4,6	94	i	
2-Chloromethoxybenzene	195				4	130	4	95	iii	
4-Chloromethoxybenzene (p-Chloroanisole)	198				2	151	2	98	iii	
1,2-Dimethoxybenzene(Veratrole)	207	22		56	4	136	4	95	v	
							4,5	132	ii	
1,3-Dimethoxybenzene	217			57	4	166	2,4,6	124	ii	
2-Bromophenyl ethyl ether (o-Bromophenetole)	218				4	134	4	98	iii	
2-Bromomethoxybenzene (o-Bromoanisole)	218				4	140	4	106	iii	
4-Bromomethoxybenzene (p-Bromoanisole)	223	12			2	148	2	88	iii	

Table 21. Ethers (*cont.*)

	B.p.	M.p.	Alkyl 3,5-dinitrobenzoate (p. 52)	Picric acid complex (p. 53)	Sulphonamide (p. 53) position	M.p.	Nitro deriv. (p. 52) position	M.p.	method	Notes
1,2-Methylenedioxy-4-(prop-2-enyl)-benzene (Safrole)	232	11		104			1,3,5	51	iii	Unsaturated; Br$_2$ in ether → pentabromo deriv., 169 (p. 53)
4-(Prop-1-enyl)methoxybenzene (Anethole)	235	21		70						Unsaturated: tribromo deriv., 108 (p. 53) CrO$_3$-ethanoic acid → 4-methoxybenzoic acid, 184 (p. 44)
Diphenyl ether	259	28		110	4,4′	159	4,4′	144	iii	
1-Methoxynaphthalene	271			130	4	156	2,4,5	128	iii	
1-Ethoxynaphthalene	280	5		119	4	164	2,4,5	149	iii	
Dibenzyl ether	295d	3	112	78						Dibromo deriv., 107 (p. 53)
1,2,3-Trimethoxybenzene	241	47		81	2,3,4	123	5	106	ii	
1,4-Dimethoxybenzene	212	56		48	2	148	2	72	i	
2-Methoxynaphthalene		72		117	8	151	1,6,8	215d	iii	

Table 22. Halides, alkyl mono-

	Chloride B.p.	Bromide B.p.	Iodide B.p.	Thiouronium picrate (p. 53) M.p.	2-Naphthyl ether (p. 54) M.p.	2-Naphthyl ether picrate (p. 54) M.p.	Notes
Methyl	−24	4	43	224	72	118	
Ethyl	12	38	72	188	37	102	
Prop-2-yl	36	60	89	196	41	95	
Prop-2-enyl	46	70	102	154	16	99	Unsaturated
Propyl	46	71	102	177	40	81	
2-Methylprop-2-yl(t-Butyl)	51	72	98	151			
1-Methylpropyl(s-Butyl)	67	90	120	166	34	86	
2-Methylpropyl(isobutyl)	68	91	120	167	33	84	
Butyl	77	101	130	177	33	67	
3-Methylbutyl (Isopentyl)	100	119	147	173	28	94	

Table 22. Halides, alkyl mono- (*cont.*)

	Chlor-ide B.p.	Bro-mide B.p.	Io-dide B.p.	Thio-uro-nium pic-rate (p. 53) M.p.	2-Naph-thyl ether (p. 54) M.p.	2-Naph-thyl ether picrate (p. 54) M.p.	Notes
Pentyl	107	128	155	154	25	66	
1-Chloro-2,3-epoxypropane (Epichlorohydrin)	116						See Table 21
Hexyl	134	156	180	157			
Cyclohexyl	142	166	180d		116		
Heptyl	160	178	203	142			
Benzyl	179	198	24*	187	102	122	*M.p.
Octyl	183	203	225	134			
2-Phenylethyl	190	218		139	70	83	
1-Phenylethyl	195	205		167			
2-Chlorobenzyl	214	102/9 mm		213			$CrO_3 \rightarrow$ 2-chlorobenzoic acid, 142 (p. 44)
3-Chlorobenzyl	215	109/10 mm		200			$CrO_3 \rightarrow$ 3-chlorobenzoic acid, 158
2-Bromobenzyl	125/20 mm	31*	47*	222			*M.p. $CrO_3 \rightarrow$ 2-bromo-benzoic acid, 150
3-Bromobenzyl	119/18 mm	41*	42*	205			*M.p. $CrO_3 \rightarrow$ 3-bromobenzoic acid, 155
4-Chlorobenzyl	214	51*	64*	194			*M.p. $CrO_3 \rightarrow$ 4-chlorobenzoic acid, 241
	M.p.	M.p.	M.p.				
3-Nitrobenzyl	45	58	84				$KMnO_4 \rightarrow$ 3-nitrobenzoic acid, 141
2-Nitrobenzyl	48	46	75				$KMnO_4 \rightarrow$ 2-nitrobenzoic acid, 146
4-Bromobenzyl	50	63	80	219			$CrO_3 \rightarrow$ 4-bromobenzoic acid, 251
4-Nitrobenzyl	71	99	127				$KMnO_4 \rightarrow$ 4-nitrobenzoic acid, 240

93

Table 23. Halides, alkyl poly-

	B.p.	Thio-uronium picrate (p. 54)	2-Naphthyl ether (p. 54)	Notes
Dichloromethane (Methylene dichloride)	42	267	133	
trans-1,2-Dichloroethene	48			Unsaturated. Dibromide, 192
cis-1,2-Dichloroethene	60			Unsaturated. Dibromide, 192
1,1-Dichloroethane (Ethylidene dichloride)	60		200	
Trichloromethane (Chloroform)	61			Forms carbylamine with pri.amines and boiling ethanolic KOH
Carbon tetrachloride	77			Forms carbylamine (as above)
1,2-Dichloroethane	83	260	217	
Trichloroethene	90			Unsaturated. Chlorine atoms are unreactive
Dibromomethane	97	267	133	
1,2-Dichloropropane	98	232	152	
1-Bromo-2-chloroethane	106		217	
1,1-Dibromoethane	112			
Tetrachloroethene	121		117	Unsaturated. Chlorine atoms are unreactive
1,3-Dichloropropane	123	229	148	
1,2-Dibromoethane (m.p. 10)	132	260	217	
1,2-Dibromopropane	141	232	152	
1,1,2,2-Tetrachloroethane	147			
Tribromomethane (Bromoform)	149			Forms carbylamine (see chloroform)
1,3-Dibromopropane	167	229	148	
Di-iodomethane	180		133	
(Dichloromethyl)benzene (Benzal chloride)	212			Conc. H_2SO_4 at 50° → benzaldehyde, which may be characterized as the 2,4-dinitrophenylhydrazone, 237
Trichloromethylbenzene (Benzotrichloride)	221			Boiling with aq. Na_2CO_3 → benzoic acid, 122
1,5-Dibromopentane	221	247		
	M.p.			
1,2-Di-iodoethane	82	260	217	
Carbon tetrabromide	91			
Tri-iodomethane (Iodoform)	119			Yellow. With quinoline in ether → compound, 65
Hexachloroethane	187*			*Sublimes. Camphor-like odour

94

Table 24. Halides, aryl

	B.p.	M.p.	Sulphonamide (p. 55) position	Sulphonamide (p. 55) M.p.	Nitro deriv. (p. 55) position	Nitro deriv. (p. 55) M.p.	method	Notes
Fluorobenzene	85		4	125	4	27	ii	
2-Fluorotoluene	114		5	105				
3-Fluorotoluene	116		6	173				
4-Fluorotoluene	117		2	141				
Chlorobenzene	132		4	143	2,4	52	ii	
Bromobenzene	156		4	162	2,4	75	ii	
2-Chlorotoluene	159		5	126	3,5	64	ii	
3-Chlorotoluene	162		6	185	4,6	91	ii	
4-Chlorotoluene	162		2	143	2,6	76	i	
1,3-Dichlorobenzene	173		4	180	4,6	103	ii	
1,2-Dichlorobenzene	179		4	135	4,5	110	ii	
2-Bromotoluene	181		5	146	3,5	82	ii	
3-Bromotoluene	184		6	168	4,6	103	i	
4-Bromotoluene	185	28	2	165	2	47	iii	
Iodobenzene	188				4	174	i	Br_2 + Fe → 4-bromo deriv., 91
2,4-Dichlorotoluene	197		5	176	3,5	104	ii	
2,6-Dichlorotoluene	199		3	204	3,5	121	ii	
3-Iodotoluene	204				6	84	i	
2-Iodotoluene	211				6	103	ii	
4-Iodotoluene	211	35						Boiling $HNO_3 \xrightarrow{\text{3 hr}}$ acid, 265
2-Chlorobenzyl chloride	214							See Table 22
4-Chlorobenzyl chloride	214	29						See Table 22
1,2,4-Trichlorobenzene	214	17	5	>200	3,5	103	ii	
1-Fluoronaphthalene	214							Picric acid deriv., 113 (p. 55)
3-Chlorobenzyl chloride	215							See Table 22
1,3-Dibromobenzene	219		4	190	4,6	117	iii	
1,2-Dibromobenzene	224		4	175	4,5	114	ii	
1-Chloronaphthalene	260		4	186	4,5	180	iii	
1-Bromonaphthalene	281		4	192	4	85	iii	Picric acid deriv., 134
3-Bromobenzyl bromide		41						See Table 22
1,4-Dichlorobenzene	174	53	2	180	2	56	iii*	*Without cooling
1,2,3-Trichlorobenzene	218	53	4	227	4	56	ii	
2-Chloronaphthalene		60	8	232	1,8	175	ii*	*6 hr. at 100°
4-Bromobenzyl bromide		63						See Table 22
1,3,5-Trichlorobenzene		63	2	212d	2	68	ii	
4-Bromochlorobenzene		67			2	72	iii*	*Without cooling
1,4-Dibromobenzene		87	2	195	2	84	iii*	*Without cooling
1,3,5-Tribromobenzene		120	2	222d	2,4	192	ii	
1,2,3,4-Tetrachlorobenzene		139	3	99				
			3,6	227				

Table 25. Hydrocarbons

	B.p.	M.p.	Sulphonamide (p. 56)	Nitro deriv. (p. 52) position	Nitro deriv. (p. 52) M.p.	method	Picric acid deriv. (p. 56)	Notes
2-Methyl-1,3-butadiene (Isoprene)	34							Unsaturated; polymerizes easily; maleic anhydride adduct, 64 (p. 55)
Pent-1-yne	40							Unsaturated. Hg deriv., 118 (p. 56)
Cyclopentadiene	40							Unsaturated. Maleic anhydride adduct, 164 (p. 55) Forms dimer b.p. 170d, m.p. 32 on standing
Penta-1,3-diene (Piperylene)	42							Unsaturated. Maleic anhydride adduct, 61 (p. 55)
Cyclopentene	44							Unsaturated
Benzene	80	6	53	1,3	90	ii		
Cyclohexane	81	6						Oxidation with fuming $HNO_3 \rightarrow$ hexanedioic acid, 153
Cyclohexene	83							Unsaturated. Conc. $HNO_3 \rightarrow$ hexanedioic acid, 153
Toluene	110		137	2,4	70	ii		
Ethylbenzene	136		109	2,4,6	37	ii		
1,4-Dimethylbenzene (p-Xylene)	137	15	147	2,3,5	139	ii		
1,3-Dimethylbenzene (m-Xylene)	139		137	2,4,6	182	ii		
Phenylethyne	140							Unsaturated. Hg deriv., 125 (p. 56)
1,2-Dimethylbenzene (o-Xylene)	144		144	4,5	71	ii		
Phenylethene (Vinylbenzene, styrene)	146							Unsaturated; polymerizes in the presence of a drop of H_2SO_4. Dibromide, 73
Isopropylbenzene (Cumene)	153		105	2,4,6	109	ii		
α-Pinene	156							Unsaturated. Dibromide, 164
Allylbenzene	157							Unsaturated. $CrO_3 \rightarrow$ benzoic acid, 122
Propylbenzene	159		107					
1,3,5-Trimethylbenzene (Mesitylene)	165		142	2,4,6	235	i		
1,2,4-Trimethylbenzene (Pseudocumene)	168		181	3,5,6	185	ii		
Dicyclopentadiene	170d	32						Unsaturated. Benzoquinone adduct, 157 (p. 55)
(+)-Limonene	176							Unsaturated; odour of lemons. Tetrabromide 104
4-(Prop-2-yl)toluene (p-Cymene)	176		115	2,3,6	118	ii		
Dipentene (Limonene)	181							Unsaturated; odour of lemons. Tetrabromide, 124
Indene	182							Unsaturated; polymerized by acid or heat. $HNO_3 \rightarrow$ phthalic acid, 195 Benzylidene deriv., 135 (p. 57)

Table 25. Hydrocarbons (*cont.*)

	B.p.	M.p.	Sulph-ona-mide (p. 56)	Nitro deriv. (p. 52) posi-tion	M.p.	meth-od	Picric acid deriv. (p. 56)	Notes
4-*t*-Butyltoluene	192	139		2	oil			
Tetrahydronaphthalene (Tetralin)	207	135		5,7	95	i		
1-Methylnaphthalene	241			4	71	iii	141	
2-Methylnaphthalene	241	37		1	81	i	115	
Diphenylmethane	262	26		2,2′,4,4′	172	ii		CrO_3—H_2SO_4 → benzo-phenone, 48
(−)-Camphene	160	51						Unsaturated; dibromide, 89
1,2-Diphenylethane (Dibenzyl)	284	52		4,4′	180	i		CrO_3—H_2SO_4 → benzoic acid, 122
Biphenyl	255	70		4,4′	234	iv		Br_2-ethanoic acid (boil for 2 hr) → 4,4′-dibromo deriv., 169
1,2,4,5-Tetramethyl-benzene (Durene)		79	155	3,6	205	ii	95	
Naphthalene		80		1	61	i	150	Odour of 'moth balls'; styphnic acid deriv., 168
Acenaphthylene		92					201	Dibromide, 121
Triphenylmethane		94		4,4′,4″	206	iii		
Acenaphthene		95		5	101	i	161	Styphnic acid deriv., 154
Phenanthrene		100					144	CrO_3-ethanoic acid → quinone, 202. Styphnic acid deriv., 142
2,3-Dimethylnaphthalene		104					124	Styphnic acid deriv., 149
Fluoranthene		110					182	Styphnic acid deriv., 151
2,6-Dimethylnaphthalene		111					143	Styphnic acid deriv., 159
Fluorene		115		2	156	i	84*	Gives blue colour with conc. H_2SO_4; styphnic acid deriv., 134
				2,7	199	ii		*Rather unstable
trans-1,2-Diphenylethene (*trans*-Stilbene)		124						Unsaturated; dibromide, 237, formed on warming with bromine. Styphnic acid deriv., 142 (p. 56)
Pyrene		150					227	Styphnic acid deriv., 191
Anthracene		217					138	Maleic anhydride adduct, 263 (p. 55). CrO_3-ethanoic acid → quinone, 286. Styphnic acid deriv., 180 (p. 56)

Table 26. Ketones (C, H and O)

	B.p.	M.p.	2,4-Di-nitro-phenyl-hydra-zone (p. 56)	Semi-carba-zone (p. 56)	4-Nitro phenyl-hydra-zone (p. 57)	Notes
Propanone (Acetone)	56		126	187	148	Monobenzylidene deriv., 42. Dibenzylidene deriv., 112 (p. 57)
Butan-2-one (Methyl ethyl ketone)	80		116	146	128	
But-3-en-2-one (Methyl vinyl ketone)	80			141		Unsaturated
Butane-2,3-dione (Diacetyl)	88		315	235 mono 278 di	230 mono >310 di	Monobenzylidene deriv., 53 (p. 57)
2-Methylbutan-3-one (Isopropyl methyl ketone)	94		124 (118)	113	108	Benzylidene deriv., 117 (p. 57)
Pentan-3-one (Diethyl ketone)	102		156	139	144	Monobenzylidene deriv., 31. Dibenzylidene deriv., 127
Pentan-2-one (Methyl propyl ketone)	102		143	110	117	
3,3-Dimethylbutan-2-one (Pinacolone)	106		125	158	139	Benzylidene deriv., 41
3-Benzoylpropanoic acid	116		191	181		See Table 16
4-Methylpentan-2-one (Isobutyl methyl ketone)	117		95	130	79	
3-Methylpentan-2-one (s-Butyl methyl ketone)	118		71	94		
2,4-Dimethylpentan-3-one (Di-isopropyl ketone)	124		96 (107)	159*		*Varies with the rate of heating
Hexan-2-one (Butyl methyl ketone)	128		106	122	88	
4-Methylpent-3-en-2-one (Mesityl oxide)	130		203	164 (133)	133*	*Prepared without heating. Unsaturated
Cyclopentanone	131		146	206	154	Benzylidene deriv., 190 (p. 57)
Methyl 2-oxopropanoate (Methyl pyruvate)	136		187	208		See Table 19
2-Methylhexan-4-one (Ethyl isobutyl ketone)	136		75	152		
2-Methylhexan-3-one (Isopropyl propyl ketone)	136		97	119		
Pentane-2,4-dione (Acetylacetone)	139		209 di 122*	107*		*Pyrazole deriv. Oxime, 149 (p. 56). See Table 18
5-Methylhexan-2-one (Isopentyl methyl ketone)	144		95	143		
Heptan-4-one (Dipropyl ketone)	144		75	133		
3-Hydroxybutan-2-one (Acetoin)	145		315	185		
Hydroxypropanone (Hydroxyacetone, Acetol)	146		129	196	173	
Heptan-2-one (Methyl pentyl ketone)	151		89	123	73	
2-Methylheptan-4-one (Isobutyl propyl ketone)	155			123		
Cyclohexanone	155		162	166	146	Benzylidene deriv., 118 (p. 57)

Table 26. Ketones (C, H and O) (*cont.*)

	B.p.	M.p.	2,4-Dinitrophenylhydrazone (p. 56)	Semicarbazone (p. 56)	4-Nitro phenylhydrazone (p. 57)	Notes
Ethyl 2-oxopropanoate (Ethyl pyruvate)	155		155	206d		See Table 19
2-Methylcyclohexanone	163		136	196	132	
4-Hydroxy-4-methylpentan-2-one (Diacetone alcohol)	165		203		209	
2-Oxopropanoic acid (Pyruvic acid)	165d		218	222	220	See Table 19
2,6-Dimethylheptan-4-one (Di-isobutyl ketone)	168		92 (66)	122		
3-Methylcyclohexanone	168		155	191 (180)	119	Benzylidene deriv., 122 (p. 57)
4-Methylcyclohexanone	169		134	198	128	Benzylidene deriv., 99 (p. 57)
Methyl 3-oxobutanoate (Methyl acetoacetate)	170		119	152		Gives red colour with aq. FeCl$_3$. See Tables 18 and 19
Octan-2-one (Hexyl methyl ketone)	173		58	122	93	
Ethyl 3-oxobutanoate (Ethyl acetoacetate)	181d		96	133	218*	Gives red colour with aq. FeCl$_3$. See Tables 18 and 19. *Pyrazole derivative
Cycloheptanone	181		148	162	137	Benzylidene deriv., 108 (p. 57)
Nonan-5-one (Dibutyl ketone)	187		41	90		
Hexane-2,5-dione (Acetonylacetone)	190		256	185 mono 220 di	115	
Nonan-2-one	194		56	120		
Fenchone	194		140	183		Unsaturated
Methyl 4-oxopentanoate (Methyl laevulinate)	196		141	144	136	
Cyclo-octanone	196		163	167		
2,6-Dimethylhepta-2,5-dien-4-one (Phorone)	198	28	112	186		
Acetophenone	202	20	249 (240)	198	184	Benzylidene deriv., 58 (p. 57)
Ethyl 4-oxopentanoate (Ethyl laevulinate)	206		101	150	157	
(−)-Menthone	207		146	184		
Decan-2-one	209	14	74	124		
Decan-3-one	211			101		
3,5,5-Trimethylcyclohex-2-en-1-one (Isophorone)	214		130	191		
1-Phenylpropan-2-one (Benzyl methyl ketone)	216	27	156	190 (196)	145	
2-Methylacetophenone (Methyl o-tolyl ketone)	216		159	206		
Propiophenone	218	18	191	174	147	
2-Hydroxyacetophenone (o-Acetylphenol)	218	28	213	210		Oxime, 117 (p. 56). See Table 30

Table 26. Ketones (C, H and O) (cont.)

	B.p.	M.p.	2,4-Di-nitro-phenyl-hydra-zone (p. 56)	Semi-carba-zone (p. 56)	4-Nitro phenyl-hydra-zone (p. 57)	Notes
3-Methylacetophenone (Methyl m-tolyl ketone)	220		207	200		
Isopropyl phenyl ketone (Isobutyrophenone)	222		163	181		
Phenyl propyl ketone (Butyrophenone)	230		190	190		
4-Methylacetophenone (Methyl p-tolyl ketone)	223		260	205		
(+)-Pulegone	224		147	174		Unsaturated
(+)-Carvone	225		190	162 (142)	174	Unsaturated
1-Phenylbutan-2-one (Benzyl ethyl ketone)	226		140	135 (146)		
Undecan-2-one	228		63	122	90	
1-Phenylbutan-3-one	235		128	142		
3-Methoxyacetophenone	240		189	196		
Butyl phenyl ketone (Valerophenone)	242		166	166	162	
2-Methoxyacetophenone	245			183		Oxime, 83
4-Oxopentanoic acid (Laevulinic acid)	245d	33	206	187d	175	Monobenzylidene deriv., 123 (p. 57). See Table 16
Diethyl 3-oxopentanedioate (Ethyl acetonedicarboxylate)	250		86	94		Red colour with aq. $FeCl_3$. See also Table 19
α-Tetralone	257		257	226	231	Benzylidene deriv., 105 (p. 57)
4-Methoxyacetophenone	258	38	227	197	195	
Methyl 1-naphthyl ketone	298	34	255	235		Benzylidene deriv., 126 (p. 57)
1,2-Diphenylpropanone (Dibenzyl ketone)	330	34	100	146		Monobenzylidene deriv., 162 (p. 57)
Benzylidenepropanone	262	41	227	186	166	Unsaturated. Benzylidene deriv., 112 (p. 57)
Indanone		42	258	233	235	Benzylidene deriv., 113 (p. 57)
Diphenyl ketone (Benzophenone)		48	238	165	154	
Methyl 2-naphthyl ketone		53	262d	236		
Phenyl 4-tolyl ketone		54 (60)	200	122		
Benzylideneacetophenone (Chalcone)		58	248 (208)	168 (180)		Unsaturated
Benzyl phenyl ketone (Deoxybenzoin)		60	204	148	163	Benzylidene deriv., 102 (p. 57)
1-Phenylbutane-1,3-dione (Benzoylacetone)		61	151		101	Monobenzylidene deriv., 99. See Table 18
4-Methoxybenzophenone		62	180 228*		199	*From trichloromethane
Fluorenone		83	284	234	269	Yellow
Benzil		95	189 mono 314 di	244d di 182 mono	290 di 192 di	Yellow
3-Hydroxyacetophenone (m-Acetylphenol)		96	256	195		See Table 30
4-Hydroxyacetophenone (p-Acetylphenol)		110	261 (225)	199		See Table 30

100

Table 26. Ketones (C, H and O) (*cont.*)

	B.p.	M.p.	2,4-Di-nitro-phenyl-hydra-zone (p. 56)	Semi-carba-zone (p. 56)	4-Nitro phenyl-hydra-zone (p. 57)	Notes
Dibenzylidenepropanone	112	180	190	172		Unsaturated
3-Benzoylpropanoic acid	116	191	181			See Table 16
2-Benzoylbenzoic acid	128					Oxime, 118; see Table 16
4-Hydroxy-3-methoxybenzylidene-propanone (Vanillideneacetone)	130	230				Unsaturated
Benzoylphenylmethanol (Benzoin)	133	245	206d			Ethanoyl deriv., 83. See Table 2
Furoin	136	216				Oxime, 161 (p. 56)
2,4-Dihydroxyacetophenone (Resacetophenone)	147	218 (244)	216			
4-Hydroxypropiophenone	148	229 240*				*From trichloromethane
Furil	162	215		199d		
2,3,4-Trihydroxyacetophenone (Gallacetophenone)	173	199	225			Oxime, 163; triethanoate, 85; see Table 30
(+)-Camphor	179	177	237	217		Benzylidene deriv., 98 (p. 57)

Table 27. Ketones (C, H, O and halogen or N)

	B.p.	M.p.	2,4-Di-nitro-phenyl-hydra-zone (p. 56)	Semi-carba-zone (p. 56)	4-Nitro-phenyl-hydra-zone (p. 57)	Notes
Chloropropanone	119		125	164*	83	*Variable
1,1-Dichloropropanone	120			163		
4-Fluoroacetophenone	196		235	219		Oxime, 80 (p. 56)
3-Chloroacetophenone	228			232	176	Oxime, 88 (p. 56)
4-Chloroacetophenone	232		236	201	239	
2-Aminoacetophenone	250d	20		290d		Oxime, 109 (p. 56). See Table 9
2-Nitroacetophenone	178/ 32 mm	27	154	210d		Oxime, 117 (p. 56)
4-Chloropropiophenone	134/ 31 mm	36	222	176		
Phenacyl bromide		50	220	146		
4-Bromoacetophenone	255	51	230	208	248	
Phenacyl chloride		59	212	156		
4-Chlorobenzophenone		78	185			Oxime, 163 (p. 56)
3-Nitroacetophenone		80	232	259		
3-Aminoacetophenone		99		196		See Table 9
4-Aminoacetophenone		106	259	250		See Table 9
4-Bromophenacyl bromide		108	218			Oxime, 115 (p. 56). Benzoic acid ester, 119
4-Phenylphenacyl bromide		126	228			Benzoic acid ester, 167
4,4'-Di-(dimethylamino)-benzophenone (Michler's ketone)		174	273			Oxime, 233 (p. 56). See Table 12

Table 28. Nitriles (some nitriles are also listed in Table 16)

	B.p.	M.p.	Carboxylic acid (p. 57)	4-Bromophenacyl ester of acid (p. 50)	Amide (p. 57)	Notes
Propenenitrile (Acrylonitrile)	78					Unsaturated. 2-Naphthol adduct, 142
Ethanenitrile (Acetonitrile)	82			85		
Propanenitrile (Propionitrile)	97			61		
2-Methylpropanenitrile (Isobutyronitrile)	108			77		
Butanenitrile (Butyronitrile)	118			63		
But-3-enenitrile (Allyl cyanide)	118			60		Unsaturated
Chloroethanenitrile (Chloroacetonitrile)	127			120		
3-Methylbutanenitrile (Isovaleronitrile)	129			68		
Pentanenitrile (Valeronitrile)	140			75		
4-Methylpentanenitrile (Isocapronitrile)	155			77		
Hexanenitrile (Capronitrile)	163			72		
2-Hydroxy-2-phenylethanenitrile (Mandelonitrile)	170d	21	118	113		
Benzonitrile	191		122	119	128	Smell of bitter almonds
2-Toluonitrile	205		104	57	142	
3-Toluonitrile	212		111	108	96	
4-Toluonitrile	218	29	180	153	160	
Propanedinitrile (Malononitrile)	219		133d		170	
Phenylethanenitrile (Benzyl cyanide)	232		76	89	157	
Hexanedinitrile (Adiponitrile)	295		153	155	220	
1-Naphthonitrile	299	35	161	135	202	
3-Chlorobenzonitrile		41	158	117	134	
2-Chlorobenzonitrile		43	142	106	142	
Butanedinitrile (Succinonitrile)		54	185	211	260	
4-Chlorobenzonitrile		92	241	126	179	
Benzene-1,2-dinitrile (Phthalonitrile)		141	200d	153	220	
4-Nitrobenzonitrile		148	240	134	201	

Table 29. Nitro-, halogenonitro-compounds and nitro-ethers

	B.p.	M.p.	position	M.p.	method	Colour with aq. NaOH	Notes
			Nitro deriv. (p. 58)				
Nitromethane	101						Acid to litmus; benzylidene deriv., 58 (p. 57)
Nitroethane	114						Benzylidene deriv., 64 (p. 57)
2-Nitropropane	120						Reduction with Sn + HCl → 2-aminopropane (p. 58)
1-Nitropropane	132						Immiscible with water. Sn + HCl → propylamine (p. 58)
Nitrobenzene	211		1,3	90	ii		Pale yellow; odour of bitter almonds. Sn + HCl → aniline (p. 58)
2-Nitrotoluene	222		2,4	70	ii		Pale yellow; odour of bitter almonds. Sn + HCl → 2-toluidine (p. 58)
1,3-Dimethyl-2-nitro benzene (2-Nitro-*m*-xylene)	226	13	2,4,6	182	ii		
Phenylnitromethane	226d						Yellow; benzylidene deriv., 75
3-Nitrotoluene	233	16					Pale yellow; Sn + HCl → 3-toluidine (p. 58) Boiling aq. $K_2Cr_2O_7$—H_2SO_4 → acid, 140 (p. 58)
6-Chloro-2-nitrotoluene	238	37					Pale yellow. $K_2Cr_2O_7$—H_2SO_4 → acid, 161 (p. 58)
1,4-Dimethyl-2-nitrobenzene (2-Nitro-*p*-xylene)	240		2,3,5	139	ii		
4-Ethylnitrobenzene	241						Sn + HCl → 4-ethylaniline (p. 58)
1,3-Dimethyl-4-nitrobenzene (4-Nitro-*m*-xylene)	244	2	2,4,6	182	ii		
2-Chloronitrobenzene	246	32	2,4	52	ii		Pale yellow
1,2-Dimethyl-3-nitrobenzene (3-Nitro-*o*-xylene)	250	15	3,4	82	ii		Pale yellow
1,2-Dimethyl-4-nitrobenzene (4-Nitro-*o*-xylene)	258	29	3,4	82	ii		
2-Methoxynitrobenzene (*o*-Nitroanisole)	265	9	2,4	88*	i		*Nitration at 0°
			2,4,6	68	ii		
2-Ethoxynitrobenzene (*o*-Nitrophenetole)	267		2,4	86*	i		*Nitration at 0°
			2,4,6	78	ii		
2-Nitrobiphenyl	320	37	2,4'	93	ii		
2-Bromonitrobenzene	259	41	2,4	72	ii		Pale yellow
2-Nitro-1,3,5-trimethyl-benzene (Nitromesitylene)	255	44	2,4	86	iv		
			2,4,6	235	ii		
3-Chloronitrobenzene	236	44	3,4	36	ii		Pale yellow; Sn + HCl → 3-chloroaniline (p. 58)
4-Nitrotoluene	234	52	2,4	70	ii		Pale yellow; odour like nitrobenzene. $K_2Cr_2O_7$-dil. H_2SO_4 → acid, 241 (p. 58)
1-Chloro-2,4-dinitrobenzene		52	2,4,6	183	ii	Red → lilac	Reactive chlorine atom; boiling 2N NaOH → 2,4-dinitrophenol, 114 Hydrazine → 2,4-dinitrophenyl-hydrazine, 199
4-Methoxynitrobenzene (*p*-Nitroanisole)		53	2,4	87	i		Boiling conc. NaOH → 4-nitro-phenol, 114

103

Table 29. Nitro-, halogenonitro-compounds and nitro-ethers (*cont.*)

	B.p.	M.p.	Nitro deriv. (p. 58) position	M.p.	meth-od	Colour with aq. NaOH	Notes
3-Bromonitrobenzene		56	3,4	59	ii		Pale yellow
1,4-Dichloro-2-nitrobenzene		56	2,6	104	ii		Pale yellow. KOH in boiling aq. methanol → 4-chloro-2-nitro-methoxybenzene, 98
1-Nitro-2-phenylethene (β-Nitrostyrene)		58					Yellow. Sn + HCl → 2-phenyl-ethylamine (p. 58)
1-Nitronaphthalene		60	1,3,8	218	ii		Yellow. Picrate, 71. CrO_3-ethanoic acid → 3-nitrobenzene-1,2-di-carboxylic acid, 218
2,6-Dinitrotoluene		66				Violet	Boiling dil. HNO_3 → acid, 202
2,4-Dinitrotoluene		70	2,4,6*	82	ii	Blue	CrO_3-conc. H_2SO_4 → acid, 183 (p. 58) *Explosive; not recommended
1,3-Dimethyl-5-nitrobenzene (5-Nitro-*m*-xylene)		75	4,5,6	125	ii		
4-Chloronitrobenzene		83	2,4	52	ii		Pale yellow. Reactive chlorine atom; boiling aq. KOH → 4-nitrophenol, 114
1-Chloro-2,4,6-trinitrobenzene (Picryl chloride)		83				Red	Yellow. Reactive chlorine atom; warm aq. KOH → picric acid, 122. Naphthalene adduct, 150
1,4-Dibromo-2-nitrobenzene		84					Pale yellow. Reactive bromine atom; boiling aq. methanolic KOH → 4-bromo-2-nitro-methoxybenzene, 86
1,3-Dinitrobenzene		90				Purple	Pale yellow. Hot ethanolic NH_4SH → 3-nitroaniline, 114 (p. 58)
1,3-Dimethyl-4,6-dinitrobenzene (4,6-Dinitro-*m*-xylene)		93	2,4,6	182	ii	Violet	Pale yellow. Hot ethanolic NH_4SH → 2,4-dimethyl-5-nitroaniline, 123 (p. 58)
1,2-Dinitrobenzene		118				None	Hot ethanolic NH_4SH → 2-nitro-aniline, 71; hot aq. NaOH → 2-nitrophenol, 45
1,4-Dinitrobenzene		172				Green-yellow	Naphthalene adduct (in ethanol), 118. Boiling ethanolic NH_4SH → 4-nitroaniline, 147 (p. 58)

Table 30. Phenols (C, H and O)

	B.p.	M.p.	FeCl₃ colour Aq.	FeCl₃ colour MeOH	Benzoate (p. 59)	Aryloxy-ethanoic acid (p. 59)	Toluene-4-sulphonate (p. 59)	3,5-Dinitrobenzoate (p. 59)	Notes
2-Methylphenol (o-Cresol)	190	31	B → G	G	Oil	152	55	138	4-Nitrobenzoate, 128 (p. 59)
2-Hydroxybenzaldehyde (Salicylaldehyde)	196		V	V		132	63		See Table 4
3-Methylphenol (m-cresol)	202	12	B → G	G	55	103	51	165	
4-Methylphenol (p-cresol)	202	35	B	YG	71	136	70	188	
2-Methoxyphenol (Guaiacol)	205	30	R	G	57	116	82	142	
2-Ethylphenol	207		B	G	38	141		108	
2,4-Dimethylphenol (1,3-Xylen-4-ol)	211	27	B	GBt	38	142		164	See Table 26
2-Hydroxyacetophenone	218	28	VR	VR	87				
3-Ethylphenol	216		V	G	52	75			
Methyl 2-hydroxybenzoate (Methyl salicylate)	224		V	V	92				4-Nitrobenzoate, 128 (p. 59) See Table 19
Ethyl 2-hydroxybenzoate (Ethyl salicylate)	233		RV	V	79 (87)				4-Nitrobenzoate, 107 (p. 59) See Table 19
4-(2-Methylpropyl)phenol	236					125			
2-Methyl-5-isopropylphenol (Carvacrol)	237			Gt		151		80	
3-Methoxyphenol	243			V		116			
4-Butylphenol	248	22			27	81			4-Nitrobenzoate, 68 (p. 59)
4-(Prop-2-enyl)-2-methoxyphenol (Eugenol)	253		YG	B	70	100*	85	131	*Hydrate, 81. Unsaturated
2-Methoxy-4-(prop-1-enyl)-phenol (Isoeugenol)	267			Gt	106	94		158	Unsaturated. Dibromide, 94
Phenol	182	42	V	G	69	99	96	146	4-Nitrobenzoate, 111 (p. 59) See Table 19
Phenyl 2-hydroxybenzoate (Salol)		42		VR	81				
4-Ethylphenol	219	47	B		60	97		132	
2,6-Dimethylphenol	203	49	Y	Y	39	140		159	
3-Methyl-6-isopropylphenol (Thymol)	233	50	—	RBr	33	148	71	103	
4-Methoxyphenol	243	54	Vt	G	87	111		166	
2-Phenylphenol		57 (67)			76	107	65		
3,5-Dihydroxytoluene, hydrate (Orcinol)		58	BV	—	88	217		190	Heating at 100° → anhyd. form, 107

Table 30. Phenols (C, H and O) (cont.)

Compound	B.p.	M.p.	FeCl₃ colour Aq.	FeCl₃ colour MeOH	Benzoate (p. 59)	Aryloxy-ethanoic acid (p. 59)	Toluene-4-sulphonate (p. 59)	3,5-Dinitro-benzoate (p. 59)	Notes
3,4-Dimethylphenol		62	B	Y	59	163		182	*Anhydrous; hydrate, 84
3,5-Dimethylphenol		68		GB	24	111*	83	195	
2,4,5-Trimethylphenol		71	—		63	132		179	4-Nitrobenzoate, 104 (p. 59)
2,5-Dimethylphenol		75	B	YG	61	118		137	
2,3-Dimethylphenol		75		G					
4-Hydroxy-3-methoxybenzaldehyde (Vanillin)		80	BV	G	78	187	115		See Table 4
4-(1,1,3,3-Tetramethylbutyl)phenol (p-t-Octylphenol)		84		G	82	188			
2-Hydroxybenzyl alcohol (Saligenin)		86			51*	108			*Dibenzoate, 85
4-(2-Methylbut-2-yl)phenol		92	Pk*	Gt	61	120	55	217	*Colour of precipitate; See Table 26
1-Naphthol		94		Br	56	192	89		
3-Hydroxyacetophenone		96			52				See Table 4
4-(2-Methylprop-2-yl)phenol (p-t-Butylphenol)		99		G	82	86	110		
3-Hydroxybenzaldehyde		104	V	—	38	148			
1,2-Dihydroxybenzene (Catechol)		105	G	G	84 (di), 131 (mono)			152	4-Nitrobenzoate, 170 (p. 59)
3,5-Dihydroxytoluene, anhyd. (Orcinol)		107	V	—	88	217		190	
4-Hydroxyacetophenone		110	V	BrR	134	177	72	138	Ethanoate, 54. See Table 26
1,3-Dihydroxybenzene (Resorcinol)		110	V	G	117 (di), 135 (mono)	195	80 (di)	201	
Ethyl 4-hydroxybenzoate		115	V	—	94				See Table 19
1,3,5-Trihydroxybenzene, dihydrate (Phloroglucinol)		117	V	G	174				
4-Hydroxybenzaldehyde		117	V	Y	91	198		162	Ethanoate, 104 (p. 59)
2,6-Dihydroxytoluene (2-Methylresorcinol)		119	VBr	Br	106				See Table 4
2-Naphthol		123	W*	Gt	107	154	125	210	*Opalescent; Ethanoate, 85. See Table 19
Methyl 4-hydroxybenzoate		131	V	—	135				
1,2,3-Trihydroxybenzene (Pyrogallol)		133	R	G	90*	198		205	*Dibenzoate, 126; monobenzoate, 138

106

Compound	m.p.							Remarks
2,4-Dihydroxybenzaldehyde (Resorcylaldehyde)	135	R	R					See Table 4
1,2,4-Trihydroxybenzene (Hydroxyquinol)	140	R*	R	120				*In the presence of a trace of aq. NaOH. Ethanoate, 96 (p. 59)
2,4-Dihydroxyacetophenone (Resacetophenone)	147	R		80 di				See Table 26
4-Hydroxypropiophenone	148							See Table 26
2-Hydroxy-5-methylbenzoic acid (5-Methylsalicylic acid)	153	VB	B	155	185			Ethanoate, 152. See Table 16
3,4-Dihydroxybenzaldehyde (Protocatechualdehyde)	154			96				See Table 4
3,5-Dihydroxybenzaldehyde	157							See Table 4
2-Hydroxybenzoic acid (Salicylic acid)	158	V	V	132	191	154		Ethanoate, 135. See Table 16
2,3-Dihydroxynaphthalene	162	B*		152				Ethanoate, 105. *Precipitate also present
2-Hydroxy-3-methylbenzoic acid (3-Methylsalicylic acid)	163	VB	V		204			Ethanoate, 113. See Table 16
4-Phenylphenol	165		G	149	190	179		
1,4-Dihydroxybenzene (Quinol, hydroquinone)	171	*		200	250	159	317	*Oxidized to 1,4-benzoquinone
2,3,4-Trihydroxyacetophenone (Gallacetophenone)	173	Br	VBr	118				Triethanoate, 85. See Table 26
1,4-Dihydroxynaphthalene	176			169				Diethanoate, 128
2,7-Dihydroxynaphthalene	186			139	149	150		Diethanoate, 136
1-Hydroxy-2-naphthoic acid	195	B	BG					Ethanoate, 158. See Table 16
3-Hydroxybenzoic acid	200		B		206	163		Ethanoate, 131. See Table 16
2,5-Dihydroxybenzoic acid (Gentisic acid)	200	BV	B	200				Diethanoate, 118; 2-ethanoate, 172; 5-ethanoate, 131. See Table 16
4-Hydroxybenzoic acid	213	O	O	221	278	169		Ethanoate, 187. See Table 16
2,4-Dihydroxybenzoic acid	213d			152				Diethanoate, 136. See Table 16
1,3,5-Trihydroxybenzene, anhydrous (Phloroglucinol)	217	V	G	174* tri		162		Triethanoate, 105. See Table 18. *Di, 126; mono, 196
3-Hydroxy-2-naphthoic acid	222			204				Ethanoate, 184. See Table 16
3,5-Dihydroxybenzoic acid	236d			227				Diethanoate, 160. See Table 16
1,5-Dihydroxynaphthalene	265	—	—	235				Diethanoate, 160

Abbreviations for colours produced by iron(III) chloride: B, blue; Br, brown; G, green; O, orange; Pk, pink; R, red; t, transient; V, violet; W, white; Y, yellow; —, no colour.

Note. In the above iron(III) chloride tests, any deviation from the solvent stated will frequently invalidate the test.

Table 31. Phenols (C, H, O and halogen or N)

	B.p.	M.p.	FeCl₃ colour Aq.	FeCl₃ colour MeOH	Benzoate (p. 59)	Aryloxy-ethanoic acid (p. 59)	Toluene-4-sulphonate (p. 59)	3,5-Dinitro-benzoate (p. 59)	Notes
2-Chlorophenol	175	7	V	V	Oil	144	74	143	
2-Bromophenol	194	5	V	V		142	78		
3-Chloro-4-methylphenol	196			G	71	108			
3-Chlorophenol	214	33			71	109			
3-Bromophenol	236	33			86	108			
2-Bromo-4-chlorophenol	123/10 mm	33			99	139	53	156	
2,4-Dibromophenol	238	36	V	YG	98	153	120		
3-Iodophenol		40			73	115	60	183	
4-Chlorophenol	217	43	BV	G	86	156	71	186	
2,4-Dichlorophenol	209	43	VB	YG	97	140	125		
2-Iodophenol		43	—		34	135	80		
2-Nitrophenol	216	45	Y	Br	59	156*	83	155	Yellow. *Difficult to purify
3-Methyl-4-nitrophenol		56			77				
4-Chloro-2-isopropyl-5-methylphenol (p-Chlorothymol)		60		Y	72			129	
4-Bromophenol		64	V	YG	102	157	94	191	
4-Chloro-3-methylphenol		66 (56)	B	YG	86	178	98	135	
2,4,6-Trichlorophenol		68	—	—	76	182	101	136	Releases CO₂ from bicarbonate
2,4,5-Trichlorophenol		68		—	93	157	115		Releases CO₂ from bicarbonate
8-Hydroxyquinoline		75	BG	V	118				4-Nitrobenzoate, 174 (p. 59) See Table 12

Compound	m.p.							Notes
1-Bromo-2-naphthol	84	—		98		121		Ethanoate, 95 (p. 59)
4,6-Dinitro-2-methylphenol	86	—		133		167		
4-Iodophenol	94		—	119		99		Releases CO_2 from bicarbonate
2,4,6-Tribromophenol	95	RV	—	81	156	113	174	Yellow
3-Nitrophenol	97	R	Br	95	200	113	159	Yellow
4-Nitrophenol	114	RBr	R	142	155	97	188	Yellow. *Difficult to purify
2,4-Dinitrophenol	114	RBr	RBr	132	184*	121		
4-Chloro-3,5-dimethylphenol	115	BG	G	68	148*	103		*N-Mono deriv., 157, O-mono deriv., 96. See Table 9
3-Aminophenol	122	Br	RBr	153 di	141	110*	179	Yellow; releases CO_2 from $NaHCO_3$; naphthalene adduct, 150 (p. 56)
2,4,6-Trinitrophenol (Picric acid)	122	—	—	163		179		See Table 14
4-Nitrosophenol	125d							
2-Hydroxybenzamide (Salicylamide)	139	R	V	143	84		224	O-Ethanoate, 138
2-Amino-4,6-dinitrophenol (Picramic acid)	169	Br	—	220 N-	191			Red. See Table 9
2-Aminophenol	174	RBr	R	184	146* (139)			*N-Mono deriv.; O-mono deriv., 101
2,4,6-Trinitroresorcinol (Styphnic acid)	179							Bright yellow. Naphthalene adduct, 168
4-Aminophenol	184d	V	V → Br 234			168 di 252 N-	178	See Table 9
Pentachlorophenol	190	—	RBr	164	196	145		Ethanoate, 150. Releases CO_2 from bicarbonate

Abbreviations for colours produced by iron(III) chloride: B, blue; Br, brown; G, green; O, orange; Pk, pink; R, red; t, transient; V, violet; W, white; Y, yellow; —, no colour.

Note. In the above iron(III) chloride tests, any deviation from the solvent stated will frequently invalidate the test.

Table 32. Quinones

	Colour	M.p.	Oxime (p. 60)	Semicarbazone (p. 60) mono	di	Quinol (p. 60)	Notes
5-Isopropyl-2-methyl-1,4-benzoquinone (Thymoquinone)	Yellow	45	162	202d	237	143	
2-Methyl-1,4-benzoquinone (p-Toluquinone)	Yellow	69	134d mono 220d di	178	240d	124	
2-Methyl-1,4-naphthoquinone	Yellow	106	166 di 160 mono	178 4-	240d	170	
1,4-Benzoquinone	Deep yellow	115	144d mono 240d di	166	243	170	
1,4-Naphthoquinone (α-Naphthoquinone)	Yellow	125	198 mono 207d di	247		176	
1,2-Naphthoquinone (β-Naphthoquinone)	Red	146d	169 di 163 2- 109 1-	184		108*	*Anhydrous; hydrate, 60
9,10-Phenanthraquinone	Orange	206	158 mono 202d di	220d		148	
Acenaphthenequinone	Yellow	261	230 mono	192	271		
9,10-Anthraquinone	Pale yellow	286	224 mono			180	4-Nitrophenylhydrazone, 238 (p. 59)

Table 33. Sulphonic acids and their derivatives

This table is arranged according to the boiling or melting point of the sulphonyl chloride because many of the acids do not have definite and reproducible values.

Sulphonyl chloride	M.p.	Acid	Amide (p. 60)	Anilide (p. 61)	Benzyl thiouronium salt of acid (p. 61)	Xanthyl deriv. of amide (p. 61)	Notes
Methane-	*	†	90	99			*B.p. 161 †B.p. 167/10 mm
Ethane-	*		58	58	115		*B.p. 177
Benzene-	14	66	155	110	148	206	Benzoyl deriv. of amide. 147; diphenylmethyl deriv., 185 (p. 62)

110

Table 33. Sulphonic acids and their derivatives (*cont.*)

Sulphonyl chloride	M.p.	Acid	Amide (p. 60)	Anilide (p. 61)	Benzyl thiouro-nium salt of acid (p. 61)	Xanthyl deriv. of amide (p. 61)	Notes
2,5-Dichlorobenzene-	38	93	181	160	170		Ethanoyl deriv. of amide, 214 (p. 61)
4-Chlorobenzene-	53	68	144	104	175		
2,4,6-Trimethylbenzene-	57	77	142	109		203	Ethanoyl deriv. of amide, 165
Naphthalene-1-	67	90	150	112	137		Benzoyl deriv. of amide, 194 (p. 61)
(+)-Camphor-10-	67	193	132	120	210		
4-Toluene-	69	92	137* 105	103	182	197	Diphenylmethyl deriv., 155 (p. 62); *anhyd.
4-Bromobenzene-	75	103	166	119	170		
Naphthalene-2-	76	91	217	132	191		Ethanoyl deriv. of amide, 145
2-Carboxybenzene-	79	68 hyd. 134 anhyd.	230*	194	206	199	*Saccharin (sulph-imide). See Table 6
Anthraquinone-2-	197		261	193	211		
Anthraquinone-1-	214	218		216	191		
4-Hydroxybenzene- (Phenol-*p*-sulphonic)			177	141	169		Warm Br$_2$-water → tri-bromophenol, 95
4-Aminobenzene- (Sulphanilic acid)		>300d	165*		182	208	Dibenzoyl deriv. of amide, 268 *See Table 10

Table 34. Thioethers (sulphides)

	B.p.	M.p.	Sulphone (p. 62)	Notes
Dimethyl thioether	38		109	
Diethyl thioether	92		73	
Dipropyl thioether	142		29	
Di-(2-methylpropyl) thioether	172		17*	*B.p. 265
Dibutyl thioether	182		44	
Methylthiophenyl	188		88	
Ethylthiophenyl	204		41	
Diphenyl thioether	295		128	
Dibenzyl thioether	150	49	150	
Di-4-tolyl thioether	158	57	158	
Di-1-naphthyl thioether		110	187	
Di-2-naphthyl thioether		151	177	

Table 35. Thiols and thiophenols

	B.p.	M.p.	2,4-Dinitro-phenyl sulphide (p. 62)	H 3-nitro-phthaloyl deriv. (p. 62)	3,5-Dinitro-benzoyl deriv. (p. 62)	Notes
Methanethiol	6		128			
Ethanethiol	36		115	149	62	
Propane-2-thiol	58		95	145	84	
Propane-1-thiol	68		81	137	52	
Prop-2-en-1-thiol	90		72			
2-Methylpropane-1-thiol	88		76	136	64	
Butane-1-thiol	98		66	144	49	
3-Methylbutane-1-thiol	117		59	145	43	
Pentane-1-thiol	127		80	132	40	
Ethane-1,2-dithiol	146		248			
Hexane-1-thiol	111		74			
Cyclohexanethiol	159		148			
2-Hydroxyethanethiol	160		101			
Benzenethiol (Thiophenol)	169		121	130	149	
Propane-1,3-dithiol	169		194			
Heptane-1-thiol	176		82	132	53	
Toluene-α-thiol	194		130	137	120	
Toluene-2-thiol (Thio-o-cresol)	194	15	101			
Toluene-3-thiol (Thio-m-cresol)	195		91			
Octane-1-thiol	199		78			
Naphthalene-1-thiol (α-Thionaphthol)	209		176			
Toluene-4-thiol	195	43	103			
4-Aminobenzenethiol (p-Aminothiophenol)		46				See Table 10
4-Chlorobenzenethiol		53	123			
Naphthalene-2-thiol (β-Thionaphthol)		81	145			

7

PHARMACEUTICAL COMPOUNDS

INTRODUCTION

Organic compounds used in medicine are so numerous and varied in character that their identification may be difficult. However, the number in common use is much smaller and a selection of these is listed in Table P1–P6. In order to characterize these pharmaceutical compounds, the following procedure is recommended.

1. Establish the elemental composition of the compound by the Lassaigne procedure (p. 1).

2. Determine the melting point or boiling point (p. 3–4).

3. By correlating these two results, it may be possible to make a tentative identification of the compound by reference to the appropriate table (P1–P6).

4. Further confirmation of the identity of the compound should be obtained from the chemical tests and spectral data given in Tables I–XIV (pp. 14–34).

5. Reference may now be made to the *British Pharmacopoeia* (or other recognized pharmacopoeia) for specific tests and spectral data which may be given for the compound.

6. Where appropriate, the compound should be converted into a crystalline derivative chosen from those given in Chapter 6.

Notes: (*a*) Some drugs exist as metallic salts of acids and others as acid salts of bases; these usually have poorly defined melting points. For such compounds, it is necessary to liberate the free acids or bases by acidification or basification, respectively. The free acid or base should be characterized in the usual way.

(*b*) For a more comprehensive list of compounds, see *Isolation and Identification of Drugs* by E. G. C. Clarke (The Pharmaceutical Press, London).

(*c*) Compounds which are also listed in the tables in Chapter 6 are indicated by the appropriate table number.

(d) Sugars are listed in Table 15 of Chapter 6.

Table P1. Compounds containing C, H, (and O)

	B.p.		M.p.
Methylpentynol	120	Phenindione	151
Eugenol (Table 30)	253	Testosterone	154
		Methyltestosterone	164
		Hydrocortisone hydrogen succinate	168
	M.p.	Stilboestrol	172
Salol (Tables 9 and 30)	42	Oestradiol	175
Butylated hydroxyanisole	62	Camphor	181
Hexylresorcinol	66	Hexoestrol	185
Vanillin (Tables 4 and 30)	80	Ascorbic acid	191d
Dimethisterone	100d	Norethisterone	206
Acetomenaphthone	112	Hydrocortisone	214d
Testosterone phenylpropionate	115	Hydrocortisone acetate	220d
Calciferol	116	Prednisolone	229d
Progesterone	128	Prednisone	230d
Aspirin (Table 16)	136	Dienoestrol	232
Cholesterol	148	Phenolphthalein	258

Table P2. Compounds containing C, H, N, (and O)

	B.p.		M.p.
Amphetamine	200	Practolol	144
Nikethamide	280	Hexobarbitone	145
		p-Aminosalicylic acid	150
	M.p.	Codeine	156
Amethocaine	44	Salbutamol	156
Lignocaine	67	Amylobarbitone	157
Hydroxyquinoline	76	Quinidine	168
Glutethimide	86	Paracetamol	170
Benzocaine (Tables 9 and 19)	90	Isoniazid	172
Oxyphenbutazone	96	Quinine	174
Pholocodine	97	Cyclobarbitone	171
Quinalbarbitone	100	Levallorphan tartrate	176
Mepyramine maleate	101	Phenobarbitone	177
Meprobamate	104	Barbitone	190
Phenylbutazone	106	Adrenaline	212d
Amidopyrine	107	Nitrazepam	229
Phenazone	111	Mefenamic acid	230
Acetanilide (Table 7)	114	Nicotinic acid (Table 17)	235
Atropine	114	Caffeine	236
Levorphanol tartrate	116	Nitrofurantoin	255
Butobarbitone	124	Theophylline	271
Bisacodyl	135	Levodopa	277d
Phenacetin (Table 7)	135	Primidone	280
Methoin	138	Methyldopa	290
Salicylamide (Tables 6 and 31)	139	Theobromine	290
Orthocaine	143	Phenytoin	296d

Table P3. Compounds containing C, H, Halogen (and O)

	B.p.		M.p.
Halothane	50	Chloroxylenol (Table 31)	115
Ethchlorvynol	174	Chlorotrianisene	118
		4-Chloro-2-methylphenoxyacetic acid	120
	M.p.	Ethacrynic acid	122
Chlorocresol	65	2,4-Dichlorophenoxyacetic acid	138
Chlorbutol	77	Fludrocortisone acetate	225
Butyl chloral hydrate	78	Betamethasone	246d
Chlorphenesin	81	Dexamethasone	255d
Dicophane	109	Fluoxymesterone	278

Table P4. Compounds containing C, H, N, Halogen (and O)

	M.p.		M.p.
Chlorambucil	67	Amylocaine hydrochloride	177
Lignocaine hydrochloride	77	Pethidine hydrochloride	189
Cetylpyridinium chloride	80	Amitriptyline hydrochloride	197
Alprenolol hydrochloride	111	Morphine hydrochloride	200d
Carbromal	120	Ephedrine hydrochloride	218
Flufenamic acid	125	Nortriptyline hydrochloride	218
Chlorpheniramine maleate	133	Procyclidine hydrochloride	227
Diazepam	134	Diamorphine hydrochloride	229
Chloramphenicol	151	Aminacrine hydrochloride	234
Procaine hydrochloride	155	Gallamine triethiodide	235d
Indomethacin	162	Methadone hydrochloride	236
Procainamide hydrochloride	167	Antazoline hydrochloride	240
Imipramine hydrochloride	170	Pyrimethamine	242
Diphenhydramine hydrochloride	170	Decamethonium iodide	246
Isoprenaline hydrochloride	172	Nalorphine hydrobromide	260d
Methylamphetamine hydrochloride	173	Acetrizoic acid	280d

Table P5. Compounds containing C, H, N, S (and O)

	M.p.		M.p.
Phenoxymethylpenicillin	124	Quinidine sulphate	200d
Carbimazole	125	Sulphathiazole	201
Isoprenaline sulphate	128d	Sulphadimethoxine	204
Tolbutamide	128	Orciprenaline sulphate	205
Poldine methylsulphate	137	Sulphamethizole	211
Sulphanilamide (Table 10)	165	Propylthiouracil	220
Dapsone	176	Saccharin (Table 6)	230
Sulphacetamide	181	Sulphamerazine	235d
Succinylsulphathiazole	189d	Procainamide sulphate	236d
Sulphaguanidine	191	Sulphadiazine	255d
Sulphapyridine	191	Acetazolamide	258
Atropine sulphate	194	Ephedrine sulphate	258d
Sulphafurazole	198	Phthalylsulphathiazole	272d
Sulphadimidine	198	Amphetamine sulphate	300d
Probenecid	199	Mercaptopurine	300d

Table P6. Compounds containing C, H, N, S, Halogen (and O)

	M.p.		M.p.
Chlorpropamide	128	Promethazine hydrochloride	223
Promazine hydrochloride	181	Bendrofluazide	225d
Chlorpromazine hydrochloride	196	Chlorothiazide	343
Chlorthalidone	220		

INDEX

Acetals,
  physical constants, 64
  preparation of derivatives, 42–43
  reactions, 15
  solubility, 5
Acyl halides,
  physical constants, 81–85
  reactions and spectral data, 26
  solubility, 6
Adsorbants for chromatography, 35, 36, 38
Alcohols,
  physical constants, 64–66
  preparation of derivatives, 43–44
  reactions and spectral data, 16
  solubility, 5
Aldehydes,
  physical constants, 67–69
  preparation of derivatives, 44–45
  reactions and spectral data, 15
  solubility, 5
Alkenes,
  physical constants, 96–97
  reactions and spectral data, 20
Alkyl halides,
  physical constants, 92–94
  preparation of derivatives, 53–54
  reactions and spectral data, 26
  solubility, 6
Alkynes,
  physical constants, 96–97
  preparation of derivatives, 56
  reactions and spectral data, 20
Alumina, chromatographic use, 35
Amides,
  physical constants, 69–70
  preparation of derivatives, 45–46
  reactions and spectral data, 21, 24
  solubility, 5
Amides, N-substituted,
  physical constants, 71
  preparation of derivatives, 46–47
  reactions and spectral data, 25
  solubility, 5
Amines,
  physical constants, 71–77
  preparation of derivatives, 46–48
  reactions and spectral data, 21–22

  solubility, 5
Amino-acids,
  physical constants, 78–79
  preparation of derivatives, 48
  reactions and spectral data, 21–23
  solubility, 5
Ammonium salts,
  reactions and spectral data, 24
  solubility, 5
Anhydrides,
  physical constants, 81–85
  reactions and spectral data, 17–18
  solubility, 5
Aryl halides
  physical constants, 95
  preparation of derivatives, 55
  reactions and spectral data, 26
  solubility, 6
Arylhydrazines,
  physical constants, 79
  reactions and spectral data, 21
  solubility, 5
Azo compounds,
  physical constants, 79
Azoxy compounds,
  physical constants, 79

Benzene ring substituents, infrared data,
  30–31
  NMR data, 10–12
Boiling point determination, 4

Carbohydrates,
  physical constants, 80
  preparation of derivatives, 48–49
  reactions and spectral data, 16
  solubility, 5
Carboxylic acids,
  physical constants, 81–87
  preparation of derivatives, 49–50
  reactions and spectral data, 14
  solubility 5
Chemical shift of protons, 10, 31–33
Chromographic columns,
  capillary, 36
  packed, 37

117

Chromatography,
  gas—liquid, 37
  high performance liquid, 36
  pyrolysis—gas, 38
  thin layer, 35—36
Chromophoric groups, 3
Colour of organic compounds, 3

Derivatives, preparation of, 42—62
Derivatives, tables of, 63—112
Detectors,
  differential refractometer, 36
  flame ionization, 37
  katharometer, 37
  ultraviolet, 36
Disaccharides,
  physical constants, 80
Drugs,
  identification, 113
  physical constants, 113—115

Enols,
  physical constants, 88
  preparation of derivatives 50
  reactions and spectral data, 14
Esters, carboxylic,
  physical constants, 88—90
  preparation of derivatives, 51—52
  reactions and spectral data, 17—18
  solubility, 5
Esters, phosphoric,
  physical constants, 90
  reactions and spectral data, 27
  solubility, 5
Ethers,
  physical constants, 91—92
  preparation of derivatives, 52—53
  reactions and spectral data, 19
  solubility, 5

Fehling's test, 16, 21
Ferrox test, 1, 2
Fingerprint region, IR, 8
Fluorolube, 8
Functional groups, tests for, 14—29

Guanidines,
  physical constants, 69—70
  preparation of derivatives, 45—46

Halides, alkyl and aryl,
  physical constants, 92—95
  preparation of derivatives, 53—55
  reactions and spectral data, 26
  solubility, 6

Halogenonitro compounds,
  physical constants, 103—104
  preparation of derivatives, 58
Halogens, tests for, 2
Hydrazines, substituted,
  physical constants, 79
  preparation of derivatives, 55
  reactions and spectral data, 21—22
Hydrocarbons,
  physical constants, 96—97
  preparation of derivatives, 55—56
  reactions and spectral data, 20, 30—31
  solubility, 5

Ignition, 3
Imides,
  physical constants, 69—70
  preparation of derivatives, 45—46
  reactions and spectral data, 24
  solubility, 5
Infrared spectroscopy,
  interpretation of spectra 8—9
  preparation of samples, 7—8

Ketones,
  physical constants, 98—101
  preparation of derivatives, 56—57
  reactions and spectral data, 15
  solubility, 5

Lactones,
  physical constants, 89
  reactions and spectral data, 17
  solubility, 5
Lassaigne's test, 1, 113

Mass spectrometry, 13
Melting point determination, 3
  mixed m.p., 4
Mixtures, separation of, 39—41
Monosaccharides,
  physical constants, 80
  preparation of derivatives, 48, 49
  reactions and spectral test, 16

Nitriles,
  physical constants, 81—85, 102
  preparation of derivatives, 57
  reactions and spectral data, 24
Nitro compounds,
  physical constants, 103—104
  preparation of derivatives, 58
  reactions and spectral data, 29—30
  solubility, 5
Nitroethers,
  physical constants, 103—104

preparation of derivatives, 58
Nitrogen, test for, 1
Nitrophenols, 26
Nitroso compounds,
    physical constants, 79
NMR spectroscopy,
    data, 31–34
    interpretation of spectra, 10–13
    preparation of samples, 10
Nujol, 8

Odour of organic compounds, 3
Organic bases, salts of,
    reactions and spectral data, 28
    solubility, 6
Oxygen, test for, 1, 2

Pharmaceutical compounds,
    identification, 113
    physical constants, 113–115
Phenols,
    physical constants, 105–109
    preparation of derivatives, 59–60
    reactions and spectral data, 14
    solubility, 6
Phosphoric acid esters,
    physical constants, 90
    reactions and spectral data, 27
    solubility, 6
Phosphoric acid salts,
    reactions and spectral data, 29
Phosphorus, test for, 2
Plates for TLC, 35
Polyols, solubility, 5
    derivatives, 65, 66

Quaternary ammonium salts,
    reactions and spectral data, 28
    solubility, 6
Quinones,
    physical constants, 110
    preparation of derivatives, 60
    reactions and spectral data, 22–23
    solubility, 5

R_f-values, 35

Schiff's reagent, 2
Schotten-Baumann reaction, 39
Silica gel, chromatographic use, 35
Siwoloboff's boiling point determination, 4
Sodium fusion test, 1
Solubility of organic compounds, 4–6

Spectroscopy,
    functional groups, 14–29
    interpretation, 10
    practical details, 7, 9, 10
    solvents for, 7(IR), 9(UV), 10(NMR)
Spray reagents, 36
Stationary phase, 35
Sulphides (thioethers),
    physical constants, 111
    preparation of derivatives, 62
    reactions and spectral data, 27
Sulphonamides,
    N-substituted, 29
    physical constants, 110–111
    preparation of derivatives, 61
    reactions and spectral data, 29
    solubility, 6
Sulphonic acids,
    physical constants, 110–111
    preparation of derivatives, 60–62
    reactions and spectral data, 27
    solubility, 5
Sulphonyl halides,
    physical constants, 111
    preparation of derivatives, 60–61
    reactions and spectral data, 28
    solubility, 6
Sulphur, test for, 2

Tests, preliminary, 1
Thioamides,
    reactions and spectral data, 29
    solubility, 6
Thioethers (sulphides),
    physical constants, 111
    preparation of derivatives, 62
    reactions and spectral data, 27
Thiols,
    physical constants, 112
    preparation of derivatives, 62
    reactions and spectral data, 27
    solubility, 5
Thiophenols,
    physical constants, 112
    preparation of derivatives, 62
    reactions and spectral data, 27
    solubility, 5
Thioureas, physical constants, 69, 70

Ultraviolet spectroscopy, 9
Ureas,
    physical constants, 69–70
    preparation of derivatives, 45–46

119